London Mathematical Society Lecture Note Series. 8

Integration and Harmonic Analysis on Compact Groups

R.E.EDWARDS

Department of Mathematics, I. A. S.
Australian National University
Canberra

Cambridge · At the University Press · 1972

CAMBRIDGE UNIVERSITY PRESS
Cambridge, New York, Melbourne, Madrid, Cape Town, Singapore, São Paulo

Cambridge University Press
The Edinburgh Building, Cambridge CB2 8RU, UK

Published in the United States of America by Cambridge University Press, New York

www.cambridge.org
Information on this title: www.cambridge.org/9780521097178

First published and Manufactured in Australia by
The Australian National University, Canberra 1970
Reprinted with corrections 1971
First Cambridge University Press edition 1972
Re-issued in this digitally printed version 2007

A catalogue record for this publication is available from the British Library

Library of Congress Catalogue Card Number: 77-190412

ISBN 978-0-521-09717-8 paperback

Contents

General Introduction

This set of notes is the result of fusing two sets of skeletal notes, one headed 'The Riesz representation theorem' and the other 'Harmonic analysis on compact groups', the aim being to end up with a reasonably self-contained introduction to portions of analysis on compact spaces and, more especially, on compact groups.

The term 'introduction' requires emphasis. These notes are not (and cannot be) expected to do much more than convey a general picture, even though a few aspects are treated in some detail. In particular, a good many proofs easily accessible in standard texts have been omitted; and many of the proofs included are presented in a somewhat condensed form and may require further attention from readers who decide to study in more detail the areas under discussion. These features arise from a deliberate attempt to avoid too much detail; they are also to some extent inevitable consequences of an attempt to survey rapidly a fairly large body of material.

The substructure of Part 2 has (I am told) been found useful as a lead-in by research students whose subsequent interest has been in specialised topics in harmonic analysis. Part 1 has, I think, filled a similar role in relation to abstract integration theory. If the readers have been attracted by the topics presented, they have pressed on to study some of the more detailed items listed in the bibliography. (In respect of Part 2, there is little doubt that the second volume of Hewitt and Ross [1] is the main follow-up to these notes.) It is hoped that the present fusion will be more helpful than either of the original sets of notes could have been when taken singly.

There are reasonable grounds for this hope, insofar as little depth in problems of harmonic analysis can be achieved without a suitable integration theory. This is the case, notwithstanding what is written in Edwards [4] to suggest that a grasp of some of the fundamental problems demands no more than a relatively primitive concept of integration; for, as was pointed out there, further pursuit of these problems usually demands a fully-fledged Lebesgue-type integration theory. Thus Part 1 goes some way to furnishing the needs of Part 2, as well as constituting the foundation for numerous other topics in abstract analysis.

To mark the end of subsections of the text, in cases where it is perhaps not otherwise obvious that the end has been reached, the 'box' symbol □ has been used.

Acknowledgements

Part 1 is a modified version of notes prepared for use in a reading course in the School of General Studies in or around 1963, and I am grateful to the students who took that course for helping to clarify the exposition. (Most of the modifications made to Part 1 are such as to make it blend more harmoniously with Part 2; for any shortcomings involved in these modifications the said students naturally are in no way responsible.)

Part 2, the major component, amounts to a considerably modified and expanded version of some skeletal notes first prepared for use at Birkbeck College, University of London, in or around 1956. Interim and relatively minor revisions to these basic notes were made in May, 1964 with the help of Dr. Garth Gaudry, and again in November, 1968 with the help of Dr. John Price. As to the present revision, Dr. Price read and criticised most ably the almost-final version of Parts 1 and 2 up to Section 2.12, together with most of the Appendices. Many of his suggestions have been incorporated in the final version as it now appears. Miss Lyn Butler contributed handsomely by checking some of the exercises.

Mr. Walter Bloom has kindly indicated a number of misprints and errors in the first printing.

To Drs Gaudry and Price, Miss Butler and Mr. Bloom I express sincere thanks. The editor, Dr. M. F. Newman, earns my gratitude for his constant readiness to discuss questions of general layout and strategy.

Finally, I am grateful to Mrs. A. Zalucki for her careful work on the typescript.

Part 1 · Integration and the Riesz Representation Theorem

1. 0. Introduction to Part 1

A hint of the flavour of abstract harmonic analysis can (as is indicated in Edwards [4]) be transmitted as soon as a relatively primitive concept of invariant integration on groups is available; for the hint to get across, it suffices that one can integrate (say) continuous functions with compact supports. In order to make a more serious study, it is necessary (as was indicated loc. cit.) to have a more highly developed integration theory of the Lebesgue type. The major aim of Part 1 is to provide a brief account of one way of extending a primitive integration theory into such a Lebesgue-type theory.

This aim might be attained in any one of several ways. The chosen method might be said to be that which fulfils most expeditiously the secondary aim of exhibiting some aspects of the general role of integration theory in functional analysis and abstract analysis in general. Since this role is to a large extent crystallised in the so-called Riesz representation theorem (RRT, for short), the selected approach to integration theory is accordingly the one which is dominated by the idea of viewing integration as a linear functional defined on a space of continuous functions. This approach is in opposition to accounts which (cf. 1. 1 below) base integration on a given measure function: instead, the measure function is made to appear as a derivative concept.

No attempt will be made to present this approach in the most general setting possible; in fact, we shall assume (except in various 'asides') that the underlying space is compact and Hausdorff. This restriction brings with it a number of technical simplifications, while yet preserving enough generality to bring out most of the important features and to ensure general utility. (Extensions and historical remarks will appear in 1. 9 and 1. 10, respectively.)

To come closer to particularities, the representation problem we intend to tackle is the following. Suppose X is a set. Denote by $B(X)$ the linear space of bounded real-valued functions on X. Let L denote a linear subspace of $B(X)$. Consider linear functionals F on L which are continuous in the sense that

$$|F(f)| \leq \text{const.}\ \|f\| = \text{const.}\ \sup_{x \in X} |f(x)|$$

for every $f \in L$. If L is finite dimensional, this continuity requirement is fulfilled by every linear functional F on L, and F is expressible as a weighted sum:

$$F(f) = \sum_{j=1}^{n} c_j f(x_j)$$

for suitably chosen $x_j \in X$ and real numbers c_j. If L is not finite dimensional, it is too much to expect that such a representation is always possible. However, one might hope that every continuous linear F will be expressible as some sort of integral.

In 1.1 we shall show quite rapidly and painlessly that this hope is justified in the shape of the Hildebrandt-Fichtenholz-Kantorovich theorem, at least for suitable choices of L. (The required concept of integration will be defined on the way.) However, for reasons which will be pinpointed in 1.1.10, this solution is not as helpful as one might wish.

It turns out that, in certain special and very important cases, it is possible to adopt a more painstaking and more constructive approach leading to a result (the Riesz representation theorem) of the desired type which is free from shortcomings of the earlier solution. The remainder of Part 1 is concerned with this more profitable approach.

1.1. Preliminaries regarding measures and integrals

Throughout 1.1, X denotes an arbitrarily given nonvoid set. The term 'set', without further qualification, means 'subset of X'. $\mathscr{P}(X)$ is a convenient symbol for denoting the set of all subsets of X. Throughout 1.1-1.10 only real-valued functions will be considered. The extension to complex-valued functions is a routine matter; see 1.11 below.

1.1.1. By an algebra of sets is meant a set $\mathscr{A}$ of subsets of X containing X itself as a member and stable under (finite) unions and under complementation. If in addition $\mathscr{A}$ contains as a member the union of any denumerable sequence of its members, $\mathscr{A}$ is termed a σ-algebra.

A set-function on $\mathscr{A}$ means simply a real-valued function whose domain is $\mathscr{A}$. (Complex-valued set-functions appear later; see 1.11.) Such a set-function μ is said to be (finitely) additive if $\mathscr{A}$ is an algebra and if

$$\mu(A \cup B) = \mu(A) + \mu(B)$$

whenever A and B are disjoint members of $\mathscr{A}$. If moreover

$$\mu\left(\bigcup_{n=1}^{\infty} A_n\right) = \sum_{n=1}^{\infty} \mu(A_n)$$

whenever (A_n) is a disjoint sequence of members of $\mathscr{A}$ whose union belongs to $\mathscr{A}$, then μ is said to be σ-additive (= countably additive, or completely additive) on $\mathscr{A}$.

A set-function μ on $\mathscr{A}$ is said to have bounded variation (BV for short) on $\mathscr{A}$ if

$$V(\mu) \equiv \sup \sum_{k=1}^{n} |\mu(A_k)|$$

is finite, the supremum being taken with respect to all finite disjoint families $(A_k)_{1 \le k \le n}$ of members of $\mathscr{A}$. This is certainly the case if μ is non-negative and additive. (Non-negative set-functions are sometimes allowed to take the value ∞, but if they do they are no longer of BV. They are barred from further discussion here.)

For brevity an additive (resp. σ-additive) set-function of BV on an algebra (resp. σ-algebra) $\mathscr{A}$ will be termed a measure (resp. σ-measure) on $\mathscr{A}$.

1.1.2. **Examples.** Since it turns out that the great virtue of the Riesz representation theorem, when compared with the results of 1.1.9 below, hinges upon the difference between additivity and σ-additivity (see 2.1.1 below), some examples exhibiting the difference are in order.

(i) If X is a finite set, it is evident that any additive set-function on any algebra $\mathscr{A}$ of subsets of X is also σ-additive. This is false for every infinite set X, however. Counter-examples can be produced by using transfinite methods, such as the Hahn-Banach theorem (if I is as in the proof of 1.9.8 below, and if c_A denotes the characteristic function of A, then $\mu: A \mapsto I(c_A)$ is additive and not σ-additive) or the use of ultrafilters (see Edwards [2], Exercises 1.23 and 1.26).

All examples of this sort may seem somewhat artificial. The next one is somewhat more natural.

(ii) For X take the real interval [0, 1]. Let $\mathscr{A}$ consist of all subsets of X which are finite unions of subintervals of X. (The intervals may be void and may contain neither, either one, or both extremities.) $\mathscr{A}$ is an algebra of subsets of X, but not a σ-algebra.

Let g be any real-valued function on X. If J is a subinterval of X with left and right extremities a and b respectively, write $\Delta g(J) = g(b) - g(a)$. Each member A of $\mathscr{A}$ can be written as a finite union of disjoint subintervals of X, say $A = J_1 \cup \ldots \cup J_n$. The sum $\sum_{k=1}^{n} \Delta g(J_k)$ can be shown to be independent of the chosen decomposition of A. A set-function μ is accordingly unambiguously defined on $\mathscr{A}$ by putting $\mu(A) = \sum_{k=1}^{n} \Delta g(J_k)$. It can be verified that μ, so defined, is additive on $\mathscr{A}$.

However, if g be suitably chosen, μ will not be σ-additive. Thus, suppose either that $\lim g(\frac{1}{n})$ does not exist, or that it exists and is different from g(0). The subinterval $J = (0, 1]$ is the union of the disjoint subintervals $J_n = (\frac{1}{n+1}, \frac{1}{n}]$ ($n = 1, 2, \ldots$). Were μ to be σ-additive, the relation

$$\mu(J) = \sum_{n=1}^{\infty} \mu(J_n)$$

would follow. Now $\mu(J) = g(1) - g(0)$, $\mu(J_n) = g(\frac{1}{n}) - g(\frac{1}{n+1})$, so that the series $\sum_{n=1}^{\infty} \mu(J_n)$ either does not converge or, if it does, converges to the sum $g(1) - \lim g(\frac{1}{n})$, which is different from $g(1) - g(0)$. Thus μ is not σ-additive.

Notice that μ will have BV if and only if g has bounded variation on [0, 1].

That $\mathscr{A}$ itself is not a σ-algebra is of no ultimate significance, for the Hahn-Banach theorem could be used to show that μ can be extended into a positive, additive set-function on the σ-algebra $\mathscr{P}(X)$.

1.1.3. The space $B_{\mathscr{A}}(X)$. Let $\mathscr{A}$ be an algebra of sets. By an $\mathscr{A}$-function (or $\mathscr{A}$-simple function) will be meant a function on X which is representable as a finite linear combination (with real coefficients) of characteristic functions c_A, where $A \in \mathscr{A}$. Herein c_A is that function on X which takes the value 1 at points belonging to A and the value zero at all remaining points of X.

The set $B(X)$ of all bounded real-valued functions on X is a Banach space, addition of functions and multiplication of a function by a real scalar being defined as usual ('pointwise operations'), and the norm being defined by

$$\|f\| = \sup_{x \in X} |f(x)| . \tag{1.1.1}$$

It is evident that $B(X)$ contains each $\mathscr{A}$-function.

$B_{\mathscr{A}}(X)$ will denote the closure in $B(X)$ of the set of all $\mathscr{A}$-functions.

It can be shown that if $\mathscr{A}$ is a σ-algebra, then $B_{\mathscr{A}}(X)$ contains precisely those $f \in B(X)$ which are $\mathscr{A}$-measurable in the sense that

$$\{x \in X : f(x) > r\} \in \mathscr{A}$$

for each real number r. In particular, if $\mathscr{A} = \mathscr{P}(X)$, $B_{\mathscr{A}}(X) = B(X)$.

1.1.5. Exercise. Prove the last two statements.

The next step is to consider the definition of the integral $\int f d\mu$ for $f \in B_{\mathscr{A}}(X)$, μ being any measure on the algebra $\mathscr{A}$.

1.1.6. Definition of $\int f d\mu$. Take first the case in which f is an $\mathscr{A}$-function. It then admits at least one (and actually many) expressions as a finite sum

$$f = \textstyle\sum_{k=1}^{n} \alpha_k \cdot c_{A_k} ,$$

the α_k being real numbers and the A_k members of $\mathscr{A}$. The additivity of μ on $\mathscr{A}$ is easily seen to ensure that the associated sums

$$\Sigma_{k=1}^{n} \alpha_k \mu(A_k)$$

are independent of the selected expression of f. The common value of these sums is, by definition, the meaning of the symbol $\int f d\mu$.

With this convention it is evident that the functional

$$f \mapsto \int f d\mu$$

is linear on the vector space of $\mathscr{A}$-functions, and that

$$\left| \int f d\mu \right| \leq V(\mu) \cdot \|f\| \qquad (1.1.2)$$

for all $\mathscr{A}$-functions f.

Take next any $f \in B_{\mathscr{A}}(X)$. Choose any sequence (f_n) of $\mathscr{A}$-functions converging uniformly to f, i.e., such that $\lim \|f-f_n\| = 0$. Such sequences do exist by virtue of the definition of $B_{\mathscr{A}}(X)$. From (1.1.2) it follows that the sequence $(\int f_n d\mu)$ is convergent (to a finite limit), and moreover that this limit is independent of the chosen sequence (f_n) converging uniformly to f. Accordingly, $\int f d\mu$ may and will be unambiguously defined to be this common limit.

1.1.7. Exercise. Verify in detail the statements made in the penultimate sentence. Check also that (1.1.2) continues to hold for any $f \in B_{\mathscr{A}}(X)$.

1.1.8. Exercise. Show that $f \mapsto \int f d\mu$ is a continuous linear functional on $B_{\mathscr{A}}(X)$, the latter regarded as a Banach space with the norm induced on it by (1.1.1).

1.1.9. The space dual to $B_{\mathscr{A}}(X)$. The result stated in Exercise 1.1.8 has a valid converse, namely: any continuous linear functional F on $B_{\mathscr{A}}(X)$ is expressible by integration with respect to some measure μ on $\mathscr{A}$.

Indeed, if this is to be the case, only one choice of the set-function μ is possible, namely

$$\mu(A) = F(c_A) . \qquad (1.1.3)$$

It remains to be shown that μ, so defined, is a measure on $\mathscr{A}$ and that

$$F(f) = \int f d\mu \qquad (1.1.4)$$

for $f \in B_{\mathscr{A}}(X)$.

It is evident from (1.1.3) and the linearity of F, that μ is additive on $\mathscr{A}$. To prove it has BV, suppose that $(A_k)_{1 \leq k \leq n}$ is a finite disjoint family of members of $\mathscr{A}$. Define the numbers $\alpha_k = \operatorname{sgn} \mu(A_k)$. Then

$$\| \Sigma_{k=1}^{n} \alpha_k c_{A_k} \| \leq 1$$

and so

$$\Sigma_{k=1}^{n} |\mu(A_k)| = F(\Sigma_{k=1}^{n} \alpha_k c_{A_k}) \leq \| F \| ,$$

by linearity of F and the standard definition

$$\| F \| = \sup \{ |F(f)| : f \quad B_{\mathscr{A}}(X), \ \ \|f\| \leq 1 \} .$$

It appears thence that $V(\mu) \leq \| F \|$, so that μ has BV. μ is therefore a measure on $\mathscr{A}$.

Now (1.1.3), combined with the linearity of F and the linearity of the integration process, shows that (1.1.4) holds when f is any $\mathscr{A}$-function. Then, by continuity of F and by continuity of the integration process (Exercise 1.1.8), (1.1.4) must continue to hold for any $f \in B_{\mathscr{A}}(X)$.

This establishes the opening statement of this section.

A simple argument shows further that

$$\| F \| = V(\mu) .$$

To sum up, one may say that formula (1. 1. 4) establishes a linear isometry $F \leftrightarrow \mu$ between the dual (= conjugate, or adjoint) space of $B_{\mathscr{A}}(X)$ and the space of all measures on $\mathscr{A}$, the latter space of measures carrying the norm defined by

$$\|\mu\| = V(\mu) .$$

This result is sometimes referred to as the Hildebrandt-Fichtenholz-Kantorovich theorem.

1. 1. 10. Some notation. We shall agree on the following notation for subsequent use.

If X is any set and k any 'object', the constant function with domain X and range $\{k\}$ will be denoted by $\underline{k}_X$, or simply by $\underline{k}$ if X is clear from the context. As a set of ordered pairs, $\underline{k}_X$ is thus $X \times \{k\}$.

If f and g are real-valued functions on X, we write $f \leq g$ (or $g \geq f$) if and only if $f(x) \leq g(x)$ for every $x \in X$. Thus, if X is a compact space, a linear functional F on $C(X)$ is non-negative (as defined immediately following 1. 2. 2 below) if and only if $F(f) \geq 0$ for every $f \in C(X)$ satisfying $f \geq \underline{0}_X$.

1. 1. 11. Exercise. Let μ be a non-negative measure on $\mathscr{P}(X)$ which is not σ-additive (see 1. 1. 2). Exhibit a monotone sequence (f_j) extracted from $B(X)$ such that $\underline{0} \leq f_j \leq \underline{1}$, $f = \lim_{j \to \infty} f_j = \underline{0}$ (the constant function zero),

$$\inf_{j \in N} \int f_j d\mu > 0 ,$$

and therefore

$$\lim_{j \to \infty} \int f_j d\mu \neq \int f d\mu$$

Remark. This example shows that passage to the limit under the integral sign is not generally permissible with integrals with respect to finitely additive measures, even though the integrands are quite well behaved and form a monotone sequence. This disagreeable situation is much improved when integrals with respect to σ-additive measures are considered; cf. Theorem 1. 5. 3 below.

1.2. Statement and discussion of Riesz's theorem.

1.2.1. Statement of the problem. A functional analytic approach to problems related to classical analysis often focuses attention on the situation in which X is a fairly simple type of topological space (rather than a structureless set) and B(X) is replaced by its subspace C(X) comprising all bounded continuous real-valued functions on X. (Notice that B(X) = C(X), if the set X is endowed with its discrete topology.) It is often of importance, and is in any case of considerable intrinsic interest, to know what the continuous linear functionals on C(X) look like.

One answer to this problem flows immediately from the substance of 1.1.9, if one applies the Hahn-Banach theorem. If F is a CLF (= continuous linear functional) on C(X), it has an extension into a CLF on the whole of B(X). So, by 1.1.9, there exists a measure μ on $\mathscr{P}$ (X) such that (1.1.4) holds for all $f \in C(X)$. See also Edwards [4], §1.

This reply is unsatisfactory from several points of view. To begin with, μ is very far from being uniquely determined by F, i.e., there will exist in general many measures μ which represent, via (1.1.4), that CLF F which is identically zero on C(X). (If X is normal, this defect can be removed by restricting μ to be 'regular', as defined below.)

A second and more important criticism stems from the preference for a representation which will be useful. As Exercise 1.1.11 indicates, integrals $\int \ldots d\mu$ with respect to arbitrary finitely-additive measures μ can behave very oddly, most of the nice theorems (like 1.5.3, 1.5.4 and 1.8.1 below) for Lebesgue-type integrals suffering spectacular breakdowns. For this reason it is natural to consider the possibility of a representation (1.1.4) in which μ is a σ-measure on some σ-algebra, $\mathscr{A}$, such that $C(X) \subseteq B_{\mathscr{A}}(X)$. This inclusion relation will obtain if (and, for the simpler examples of X, only if) $\mathscr{A}$ contains all the so-called Borel subsets of X; see 1.9. The Borel subsets of X are, by definition, precisely the elements of the smallest σ-algebra of subsets of X which contains all open (or all closed) subsets of X.

The suggestion is therefore that we define the term Borel measure (on X) to mean a σ-measure with domain the set of all Borel sets in X, and then ask whether formula (1.1.4) holds with μ a suitably chosen Borel measure. [Note - The term 'Borel measure' is often applied, even when μ is not of BV, but this extended meaning will not be used here.]

Granted such a representation, the uniqueness of μ for a given F would not be ensured unless one imposed the extra condition that μ be regular in the sense that, for any Borel set $A \subseteq X$ and any $\varepsilon > 0$, there exists an open set $U \supseteq A$ such that $|\mu(B) - \mu(A)| \le \varepsilon$ for every Borel set B satisfying $A \subseteq B \subseteq U$.

The statement of what has come to be known briefly as the Riesz representation theorem (hereinafter referred to even more briefly as the RRT) is just a summary of the desiderata outlined above, together with hypotheses on X which suffice to render the goal attainable.

1.2.2. Theorem. Let X be a Hausdorff compact space and F any CLF on C(X). Then there exists a regular Borel measure μ on X such that (1.1.4) holds for $f \in C(X)$. Moreover, μ is uniquely determined by F, and $V(\mu) = \|F\|$.

If the existence of μ be assumed, it is not difficult to show that μ is non-negative whenever F is non-negative in the sense that $F(f) \ge 0$ for every $f \in C(X)$ satisfying $f \ge 0_X$. (Incidentally, many writers use the term 'positive' in place of 'non-negative' in this connection. For obvious reasons, either choice is regrettable, the second slightly less so than the first. A better term would be '(monotone) nondecreasing', but this is non-standard.) The converse is trivial. Since it may be shown (see Exercise 1.2.6 (b) below) that any CLF F on C(X) is expressible as the difference of two (necessarily continuous) non-negative linear functionals on C(X), an equivalent formulation of the existence part of the RRT appears in

1.2.3. Theorem. If X is any Hausdorff compact space, any non-negative linear functional F on C(X) is representable in the form (1.1.4), where μ is a non-negative regular Borel measure on X. (As before, μ is uniquely determined by F, etc.)

In Theorem 1.2.3 it is unnecessary to postulate continuity of F since, as is easily seen, this is a consequence of non-negativity; see Exercise 1.2.6 (b).

The ensuing programme is confined to providing a proof of the existence statement in Theorem 1. 2. 3. The missing discussion of uniqueness is quite simple in the case of Theorem 1. 2. 3, but less so in the case of Theorem 1. 2. 2; for details, see Hewitt and Stromberg [1], (20. 45).

Discussion of the necessity of the hypotheses placed upon X in Theorem 1. 2. 3 and of its rich background, is deferred until 1. 9. In spite of all that has been said thus far, it will appear in 1. 9. 2 that the use of σ-measures and their integrals is neither necessary to, nor always maximally convenient for, the formulation of a desirable representation theorem.

1. 2. 4. Exercise. Formulate and prove the RRT for the case in which X is a finite set, regarded as a compact space with the discrete topology.

1. 2. 5. Exercise. Let X be an infinite set. Denote by $B_0(X)$ the subspace of B(X) comprising those functions $f \in B(X)$ which 'tend to zero at infinity', i. e., have the property that, for each $\varepsilon > 0$, the set $\{x \in X : |f(x)| > \varepsilon\}$ is finite; this property is often expressed by the formula $\lim_{x \in X,\ x \to \infty} f(x) = 0$. Use the RRT to deduce that to each CLF F on $B_0(X)$ corresponds a function α on X such that

$$\sum_{x \in X} |\alpha(x)| \equiv \sup \{\sum_{x \in S} |\alpha(x)| : S \subseteq X,\ \ S \text{ finite}\}$$

is finite, and

$$F(f) = \sum_{x \in X} \alpha(x) f(x)$$

for $f \in B_0(X)$. The sum appearing here is defined to be the unique real number s with the following property - Given $\varepsilon > 0$, there exists a finite subset $S(\varepsilon)$ of X such that

$$|\sum_{x \in S} \alpha(x) f(x) - s| < \varepsilon$$

for all finite subsets S of X which contain $S(\varepsilon)$.

[Hints: Form a topological space X' as follows. As a set X' is obtained by adjoining to X one more point, say ∞. The topology on X' is specified by assigning to each $x \in X$ a neighbourhood base formed of the one set $\{x\}$, and to ∞ a neighbourhood base formed of sets $X'\backslash S$, where S ranges over the finite subsets of X. Verify that X' is Hausdorff and compact, and that all subsets of X' are Borel subsets. Show that $B_0(X)$ may be identified with the set of $g \in C(X')$ which vanish at the point ∞. Apply the RRT to $C(X')$.]

1.2.6. Exercise. (a) Show that any NNLF (= non-negative linear functional) F on $C(X)$ is continuous.

(b) Let F be a CLF on $C(X)$, as in 1.2.2. For $f \in C(X)$ satisfying $f \geq \underline{0}$ write

$$F^+(f) = \sup\{F(g) : g \in C(X),\ \underline{0} \leq g \leq f\}.$$

Prove that F^+ is additive and positive-homogeneous and can be extended into NNLF on $C(X)$. Show that $F^- = F^+ - F$ is also a NNLF on $C(X)$, so that the formula $F = F^+ - F^-$ expresses F as the difference of two NNLFs on $C(X)$. (This is the minimal decomposition of F of this sort: if $F = I - J$, where I and J are NNLFs on $C(X)$, then $I - F^+$ and $J - F^-$ are both NNLFs on $C(X)$.)

1.2.7. Remark. The result of Exercise 1.2.5 may quite easily be obtained without using the RRT, but it is instructive to see that it is subsumed as a corollary thereof. The reader may like to reverse the procedure and thus give a proof of the RRT for the special case in question.

1.3. Method of proof of RRT; preliminaries

1.3.1. One starts with a definite NNLF on $C(X)$. The aim being to express this functional as an integral, it will be denoted by I. The process is a constructive extension of I to a certain set of functions on X, to be termed those which are 'integrable' (for, or relative to, I). It is crucial that this set of integrable functions shall possess certain structural properties and, at the same time, shall include the characteristic functions of all sets belonging to some σ-algebra $\mathscr{A}$ for which $C(X) \subseteq B_{\mathscr{A}}(X)$. The σ-algebra of Borel sets certainly satisfies this demand, but is in some cases unnecessarily large; see 1.9. It is also crucial that the extension of I shall satisfy certain important 'convergence theorems'. Assuming that I has been suitably extended, the associated σ-measure μ will be obtained by defining $\mu(A) = I(c_A)$.

The procedure to be followed is by no means the shortest possible. Granted some familiarity with standard measure-cum-integration theory, one of the speediest approaches is sketched by Hewitt [4]. Here, however, no such prior knowledge is assumed, and the approach will furnish en route some of the most important properties of the integrals involved. It is in fact suitable for a self-contained theory of integration (see Bourbaki [1]).

Before embarking on the construction, which will be broken into four main stages, a few facts from general topology will be recalled.

1.3.2. If X is any topological space, $\Phi = \Phi(X)$ will denote the set of all non-negative lower semicontinuous functions on X. (ϕ is lower semicontinuous on X if and only if for each real number r, the set $\{x \in X : \phi(x) > r\}$ is open in X.)

1.3.3. It will be convenient to permit functions in Φ to take the value ∞ at some or all points. This will not lead to trouble, since the symbol $(\infty) - (\infty)$ will never appear. But it is necessary to agree upon the following conventions:

$$0.\infty = 0, \qquad c.\infty = \infty \qquad \text{if } c > 0,$$
$$(\infty) + (\infty) = \infty = c + (\infty) \qquad \text{if } c \text{ is real},$$
$$c < \infty \qquad \text{if } c \text{ is real},$$

together with commutativity in the above operations.

1.3.4. If H is a nonvoid set of functions on X, its upper envelope is the function f defined by

$$f(x) = \sup\{h(x) : h \in H\},$$

which may be ∞ at some or all points of X. This upper envelope is denoted by $\sup\{h : h \in H\}$ or $\sup_{h \in H} h$.

If $H \subseteq \Phi$, the upper envelope of H belongs to Φ.

1.3.5. Exercise. Prove the last statement. □

If ϕ and ϕ' belong to Φ, and if c is a positive number, then $\phi + \phi'$ and $c\phi$ belong to Φ. If $A \subseteq X$, then $c_A \in \Phi$ if and only if A is open in X.

Use will be made of the fact that any compact Hausdorff space X is completely regular: this means that, given $x \in X$ and any neighbourhood N of x, there exists a continuous function f mapping X into $[0, 1]$ such that $f(x) = 1$ and $f(X \backslash N) \subseteq \{0\}$. For this see Kelley [1], p. 117 and p. 147. (A completely regular compact space need not be Hausdorff.)

An important corollary of complete regularity is incorporated in

1.3.6. Exercise. If X is compact and Hausdorff, each $\phi \in \Phi$ is the upper envelope of the set of non-negative functions $f \in C(X)$ satisfying $f \leq \phi$. □

Another necessary fact is embodied in the following theorem.

1.3.7. Dini's theorem. Suppose that X is a compact space, that $f \in C(X)$, and that $H \subseteq C(X)$. Suppose also that

(a) $\sup\{h(x) : h \in H\} \geq f(x)$ for each $x \in X$;

and (b) given $h', h'' \in H$, there exists $h \in H$ such that $h \geq \sup(h', h'')$.

Then, given $\varepsilon > 0$, there exists $h \in H$ such that $h \geq f - \varepsilon$.

1.3.8. Exercise. Prove Dini's theorem by considering the open sets

$$A_h = \{x \in X : h(x) > f(x) - \varepsilon\} \ . \square$$

The first stage of the construction can now begin. From this point until the end of 1.8, the notation is fixed as follows:

X is a Hausdorff compact space;

$C = C(X)$, $\Phi = \Phi(X)$;

I is a given NNLF on $C(X)$.

1.4. First stage of extension of I

In this section the domain of I is stretched to include Φ.

1.4.1. Definition. If $\phi \in \Phi$ one defines

$$I^*(\phi) = \sup\{I(f) : f \in C,\ f \leq \phi\}.$$

According to this definition, $I^*(\phi)$ may be ∞, but it is certainly finite if ϕ is bounded. (Why?)

The non-negativity of I ensures the truth of the following statements -

(1) $I^*(f) = I(f)$ if $f \in C$, $f \geq \underline{0}$.

(2) $0 \leq I^*(\phi) \leq I^*(\phi')$ if $\phi, \phi' \in \Phi$ and $\phi \leq \phi'$.

It is also very simple to show that

(3) $I^*(c\phi) = cI^*(\phi)$ if $\phi \in \Phi$ and $c \geq 0$.

The next property is much less evident.

1.4.2. Proposition. Suppose $\phi \in \Phi$ is the upper envelope of a subset H of Φ with the property that H contains a common majorant of any two of its elements. Then

$$I^*(\phi) = \sup_{h \in H} I^*(h).$$

Proof. That $\sup I^*(h) \leq I^*(\phi)$ follows from 1.4.1(2) above. It remains to show that

$$I^*(\phi) \leq \sup I^*(h),$$

for which it suffices to show that, if $f \in C$ and $f \leq \phi$, then

$$I(f) \leq \sup I^*(h).$$

To prove this, consider the set H' of all $g \in C$ such that $g \leq h$ for some $h \in H$ (h may depend on g). It is then easy to check that H' contains a common majorant of any two of its members, and that its upper envelope is $\geq f$. By Dini's theorem 1.3.7, therefore, given

$\varepsilon > 0$, one can find $g_0 \in H'$ such that $g_0 \geq f - \varepsilon$. Hence

$$I(g_0) \geq I(f) - \varepsilon I(\underline{1}) .$$

Since $g_0 \leq h_0$ for some $h_0 \in H$, $I(g_0) \leq I^*(h_0) \leq \sup I^*(h)$. So

$$I(f) \leq \sup I^*(h) + \varepsilon I(\underline{1}) .$$

Since $\varepsilon > 0$ is arbitrarily small, the proof is finished.

Remark. The proposition applies if H comprises the terms of a monotone increasing sequence (ϕ_n) of elements of Φ. But it is remarkable that no countability restriction is placed on H. The proposition is hopelessly false for more general functions, unless one is tied down to sequences; cf. 1. 5. 3 below.

1. 4. 3. Corollary. I^* <u>is additive on</u> Φ, <u>i. e.</u>, <u>if</u> $\phi_1, \phi_2 \in \Phi$ <u>then</u>

$$I^*(\phi_1 + \phi_2) = I^*(\phi_1) + I^*(\phi_2) .$$

Proof. Let H_i $(i = 1, 2)$ be the set of non-negative functions in C which minorise (i. e., which are majorised by) ϕ_i, and let H be the set of sums $f_1 + f_2$, where $f_i \in H_i$. H satisfies the hypotheses of the above proposition, and by 1. 3. 6 its upper envelope is $\phi_1 + \phi_2$. Thus Proposition 1. 4. 2 gives

$$\begin{aligned} I^*(\phi_1 + \phi_2) &= \sup_{h \in H} I^*(h) \\ &= \sup \{I^*(f_1 + f_2) : f_1 \in H_1, f_2 \in H_2\} . \end{aligned}$$

Since f_1, f_2 are non-negative and continuous, the same is true of $f_1 + f_2$, so that

$$I^*(f_1 + f_2) = I(f_1 + f_2) = I(f_1) + I(f_2) .$$

Hence

$$\begin{aligned} I^*(\phi_1 + \phi_2) &= \sup \{I(f_1) + I(f_2) : f_1 \in H_1, f_2 \in H_2\} \\ &= \sup I(f_1) + \sup I(f_2) \\ &= I^*(\phi_1) + I^*(\phi_2) , \end{aligned}$$

the last step by very definition of I^*.

1.4.4. Exercise. Take points $x_n \in X$ and numbers $c_n \geq 0$ $(n = 1, 2, \ldots)$. Show that the formula

$$I(f) = \sum_{n=1}^{\infty} c_n f(x_n) \qquad (f \in C)$$

defines a NNLF on C if $\sum_{n=1}^{\infty} c_n > \infty$. This I is then denoted by $\sum_{n=1}^{\infty} c_n \varepsilon_{x_n}$, ε_x being the 'Dirac measure at x'. Prove that

$$I^*(\phi) = \sum_{n=1}^{\infty} c_n \phi(x_n)$$

for each $\phi \in \Phi$.

1.5. Second stage of extension of I

Here we define an extension I^{**} of I^* whose domain includes all non-negative functions on X. Since in general there will exist functions which are finite-valued and which admit no finite-valued lower semi-continuous majorant, one now sees why it is convenient to allow functions in Φ to take the value ∞. However, having done this, one may as well admit into the discussion non-negative functions taking the value ∞.

1.5.1. Definition. Let f be any non-negative function on X, finite-valued or not. One defines

$$I^{**}(f) = \inf\{I^*(\phi) : \phi \in \Phi, \quad \phi \geq f\}.$$

As in the case of I^*, a number of properties are evident.

(1) $I^{**}(\phi) = I^*(\phi)$ if $\phi \in \Phi$.

(2) $0 \leq I^{**}(f) \leq I^{**}(f')$ if $\underline{0} \leq f \leq f'$.

(3) $I^{**}(cf) = cI^{**}(f)$ if c is a number ≥ 0.

1.5.2. Exercise. By using Corollary 1.4.3, show that

$$I^{**}(f_1 + f_2) \leq I^{**}(f_1) + I^{**}(f_2),$$

i.e., I^{**} is subadditive. (It is not additive, except in certain trivial cases.) □

One of the most fundamental results concerning Lebesgue-type integrals with respect to σ-additive measures now appears in the following disguise.

1.5.3. **Theorem (Monotone Convergence).** Suppose f is the limit of a monotone increasing sequence (f_n) of non-negative functions. Then

$$I^{**}(f) = \lim_{n\to\infty} I^{**}(f_n)$$

Proof. From 1.5.1(2) above it is evident that

$$I^{**}(f) \geq \lim I^{**}(f_n) .$$

In proving the reverse one may clearly assume that $I^{**}(f_n) < \infty$ for each n.

It will be shown that for any $\varepsilon > 0$ one may choose an increasing sequence ϕ_n from Φ such that $f_n < \phi_n$ and $I^*(\phi_n) < I^{**}(f_n) + \varepsilon$. If this be done, $\phi = \lim \phi_n$ will belong to Φ, it will majorise f, and Corollary 1.4.3 will show that

$$I^*(\phi) = \lim I^*(\phi_n) \leq \lim I^{**}(f_n) + \varepsilon .$$

Accordingly $I^{**}(f) \leq I^*(\phi) \leq \lim I^{**}(f_n) + \varepsilon$. Since ε is arbitrarily small, the proof will be complete.

To construct the ϕ_n one begins by choosing ϕ'_n in Φ majorising f_n and such that $I^{**}(f_n) \leq I^*(\phi') \leq I^{**}(f_n) + \varepsilon/2^n$, and then shows that the $\phi_n = \sup_{1\leq m\leq n} \phi'_m$ satisfy the demands. Evidently, $\phi_n \in \Phi$ and $\phi_n \geq f_n$. The final step is to prove by induction on n that

$$I^*(\phi_n) \leq I^{**}(f_n) + \varepsilon(1 - 2^{-n}) . \qquad \text{(a)}$$

Now this is true for $n = 1$. Assume it true for $n = k$. One has

$$\phi_{k+1} = \sup(\phi_k, \phi'_{k+1}) . \qquad \text{(b)}$$

Since $\phi_k \geq f_k$ and $\phi'_{k+1} \geq f_{k+1} \geq f_k$, so

$$\inf(\phi_k, \phi'_{k+1}) \geq f_k . \qquad \text{(c)}$$

Also

$$\inf(\phi_k, \phi'_{k+1}) + \sup(\phi_k, \phi'_{k+1}) = \phi_k + \phi'_{k+1} . \qquad \text{(d)}$$

Using the additivity of I* (Corollary 1. 4. 3), it follows from (b), (c) and (d) that

$$I^*(\phi_k) + I^*(\phi'_{k+1}) \geq I^{**}(f_k) + I^*(\phi_{k+1}) .$$

So, by inductive hypothesis,

$$\begin{aligned} I^*(\phi_{k+1}) &\leq I^*(\phi_k) + I^*(\phi'_{k+1}) - I^{**}(f_k) \\ &\leq I^{**}(f_k) + \varepsilon(1 - 2^{-k}) + I^{**}(f_{k+1}) + \varepsilon 2^{-k-1} - I^{**}(f_k) \\ &= I^{**}(f_{k+1}) + \varepsilon(1 - 2^{-k-1}) , \end{aligned}$$

which is (a) for $n = k + 1$. The proof ends by appeal to the induction principle.

1. 5. 4. **Exercise (Fatou's lemma).** Let (f_n) be an arbitrary sequence of non-negative functions. Show that

$$I^{**}(\liminf f_n) \leq \liminf I^{**}(f_n) .$$

[Hint: Consider the functions

$$F_m = \inf_{n \geq m} f_n ,$$

noting that $F_m \uparrow \liminf f_n$. Fatou's lemma will not be needed for the proof of Theorem 1. 2. 3, but it is relevant to the discussion in 1. 9 and to Part 2.]

1. 6. The space of integrable functions

The so-called integrable functions are distinguished in a fashion which is somewhat similar to the process which leads to real numbers by completing the metric space of rationals.

First one defines the appropriate metric - or, rather, semi-metric ... space.

1.6.1. Definition. $\mathscr{F}$ will denote the set of all $f \in \mathbf{R}^X$ (= the set of all real-valued functions on X) for which

$$N(f) \equiv I^{**}(|f|) < \infty .$$

By 1.5.2 and 1.5.1(3), $\mathscr{F}$ is a linear subspace of $\mathbf{R}^X$. Also, N is a seminorm on $\mathscr{F}$ (i.e., has all the properties of a norm save that $N(f) = 0$ does not in general imply that $f = \underline{0}$). The corresponding semimetric on $\mathscr{F}$ is $(f, g) \mapsto N(f - g)$. Evidently, $C \subseteq \mathscr{F}$.

Instead of talking about an abstract completion of C, we may and will remain within the set of real-valued functions on X by taking the closure of C in $\mathscr{F}$ and verify that this is a completion of C. This motivates the following definition.

1.6.2. Definition. $\mathscr{L}$ will denote the closure in $\mathscr{F}$, relative to the seminorm N, of C. The members of $\mathscr{L}$ are termed integrable functions.

Since $\mathscr{L}$ is the closure of a linear subspace of $\mathbf{R}^X$ it is itself a linear subspace of $\mathbf{R}^X$.

If $f, g \in C$ we have

$$|I(f)-I(g)| = |I(f-g)| \leq I(|f-g|) = I^{**}(|f-g|) = N(f-g) .$$

This shows that $f \mapsto I(f)$ is a uniformly continuous linear functional on the dense subspace C of $\mathscr{L}$. It therefore has a unique continuous extension into a continuous linear functional on $\mathscr{L}$. This extension is denoted by I. In more concrete terms, if $f \in \mathscr{L}$, $I(f)$ is the common limit of $(I(f_n))$ for all sequences (f_n) extracted from C and satisfying $N(f_n - f) \to 0$ $(n \to \infty)$.

1.6.3. Exercise. Show that if f and g are non-negative elements of $\mathscr{F}$, then

$$|I^{**}(f) - I^{**}(g)| \leq N(f - g) .$$

Deduce that

$$I^{**}(f) = I(f)$$

whenever $f \in \mathscr{L}$ and $f \geq 0$.
[Hint: Note that $f \leq g + |f - g|$.]

1.6.4. Exercise. Show that if $\phi \in \Phi$, then $\phi \in \mathscr{L}$ if and only if ϕ is finite-valued and $I^*(\phi) < \infty$. In particular, any bounded $\phi \in \Phi$ belongs to $\mathscr{L}$. (Concerning the restriction that ϕ be finite-valued, see the final paragraph of 1.6.7.)

1.6.5. Exercise. Show that if $f \in \mathscr{L}$, then $|f| \in \mathscr{L}$.

Deduce that if f and g are integrable, so too are $\sup(f, g)$, and $\inf(f, g)$.
[Hints: Note that

$$\big|\, |f| - |g| \,\big| \leq |f - g| ,$$

that

$$\sup(f, g) = (f + g + |f-g|) ,$$

and that

$$\inf(f, g) = -\sup(-f, -g).$$] □

The basic convergence theorem can now be established.

1.6.6. Theorem. *Let (f_n) be a monotone increasing sequence of non-negative integrable functions which converges to a limit $f \in \mathbf{R}^X$. Then f is integrable if and only if $\lim_n I(f_n) < \infty$, in which case*

$$I(f) = \lim_n I(f_n) .$$

Proof. In any case $f_n \leq f$ for every n; if f is integrable, 1.5.1(2) and 1.6.3 give $I(f_n) \leq I(f)$ for all n, hence $\lim_n I(f_n) < \infty$.

Suppose on the other hand that $\lim_n I(f_n) < \infty$. By Theorem 1.5.3,

$$I^{**}(f) = \lim I(f_n) < \infty ,$$

i.e., $f \in \mathscr{F}$. Moreover, by the same theorem,

$$N(f-f_n) = I^{**}(f-f_n) = \lim_m I^{**}(f_m - f_n) .$$

But Exercise 1. 6. 3 shows that, if $m \geq n$, then

$$I^{**}(f_m - f_n) = I(f_m - f_n) = I(f_m) - I(f_n),$$

which tends to zero as $m, n \to \infty$. It follows that, given $\varepsilon > 0$, $N(f - f_n) \leq \varepsilon$ for all sufficiently large n. Accordingly, f belongs to the closure in $\mathscr{F}$ of $\mathscr{L}$, and so, since $\mathscr{L}$ is closed in $\mathscr{F}$, $f \in \mathscr{L}$. □

Remarks. It is easy to relax the hypothesis of non-negativity of the f_n to the demand that at least one of the f_n majorises an integrable function. (If $f_{n_0} \geq u \in \mathscr{L}$, apply Theorem 1. 6. 6 to the sequence $n \mapsto f_{n_0+n} + |u|$.)

1. 6. 7. The spaces $\mathscr{L}^p$ and L^p. Mainly to cater for the needs of Part 2, it is necessary to define these spaces for every $p \in [1, \infty]$. The form of the definition depends upon whether p is or is not finite.

Suppose first that $p \neq \infty$. In this case we mimic 1. 6. 1 and 1. 6. 2 by first replacing N by N_p, where

$$N_p(f) = (I^{**}(|f|^p))^{1/p},$$

and then following the recipe in 1. 6. 2. This leads to $\mathscr{F}^p$ and thence to $\mathscr{L}^p$. Elements of $\mathscr{L}^p$ are the so-called p^{th} power integrable functions on X. (Notice that $N_1 = N$, $\mathscr{F}^1 = \mathscr{F}$ and $\mathscr{L}^1 = \mathscr{L}$

In case $p = \infty$, the procedure has to be varied. First we agree to say that an $f \in \mathscr{F}$ (resp. a set $E \subseteq X$) is negligible (or null, or of measure zero) if and only if $N(f) = 0$ (resp. $N(c_E) = 0$); the term 'of measure zero' becomes more appropriate in appearance in 1. 7. 6 below. For the moment, the sole use of the concept is in the definition of N_∞, namely: $N_\infty(f)$ is the infimum of $m \in [0, \infty) \cup \{\infty\}$ having the property that $\{x \in X : |f(x)| > m\}$ is negligible. We then define

$$\mathscr{F}^\infty = \{f \in \mathscr{F} : N_\infty(f) < \infty\},$$

$$\mathscr{L}^\infty = \{f \in \mathscr{L} : N_\infty(f) < \infty\}.$$

(In 1. 7. 10 it will appear that, in the above definition of $\mathscr{L}^\infty$, the condition expressed by '$f \in \mathscr{L}$' could be replaced by 'f is measurable'.) Only in degenerate cases is $\mathscr{L}^\infty$ the same as the N_∞-closure in $\mathscr{F}^\infty$ of C.

True (but not so obvious as in the case $p = 1$) is the claim that, for any p, N_p is a seminorm on $\mathscr{L}^p$; this is the substance of Minkowski's inequality. It is in general not a norm; in fact, $N_p(f) = 0$ if and only if f is negligible, which in general does not imply that f vanishes everywhere, but merely outside a set which is negligible. It is also the case that $\mathscr{L}^p \subseteq \mathscr{L}^q$ whenever $p \geq q$.

$\mathscr{L}^p$ is complete relative to the seminorm N_p; the case $p = \infty$ is easy, and the case $p < \infty$ is discussed in Exercise 1. 7. 13. This is one of the most important general properties of the space $\mathscr{L}^p$.

(The preceding facts will be found established in almost any book on Lebesgue-type integration; see, for example, Edwards [2], Chapter 4.)

If we denote by M the set of negligible functions, then M is a linear subspace of $\mathscr{L}^p$ for every p. The quotient space $\mathscr{L}^p/M$ is denoted by L^p and the norm on L^p obtained from the seminorm N_p on $\mathscr{L}^p$ is usually denoted by $\|.\|_p$. The completeness of $\mathscr{L}^p$ then ensures the cardinal fact that L^p is complete relative to the norm just defined; it is therefore a Banach space. Moreover, $L^p \subseteq L^q$ whenever $p \geq q$.

A customary caution has to be issued here. The distinction between $\mathscr{L}^p$ and L^p is very often ignored (or regarded as being covered by an automatic mental adjustment on the reader's part); we shall in fact be guilty of this confusion in due course. The difference which is being glossed over is that between an individual function (element of $\mathscr{L}^p$) and the equivalence class of functions (element of L^p) corresponding to it. While the abuse is rife, it is perhaps best for the reader to become accustomed to it; all one may ask is that he be on his guard and be prepared to guarantee its harmlessness in each instance in which he lets it pass. (In particular, he should be ready to do precisely this at various places in Part 2 to follow.)

There is yet another occasional abuse which should be mentioned, namely, the common practice of including in $\mathscr{L}^p$ (and into $\mathscr{F}$) those extended real-valued functions which agree, except perhaps at the points of a negligible set, with an element of $\mathscr{L}^p$ (or $\mathscr{F}$) as defined above. This, too, is usually harmless and often very convenient. However, if this practice is adopted, there is a price to pay: the enlarged sets of functions are in general no longer linear spaces in any natural way. Without wishing to expand upon this, our advice to the reader must be much as before.

1. 7. The σ-measure associated with I; proof of the RRT

1. 7. 1. Measurable sets. A set $A \subseteq X$ is said to be measurable (though 'integrable' would be a more sensible term) if and only if c_A is integrable, in which case one writes $\mu(A) = I(c_A)$. (This concept of measurability naturally depends on the NNLF I with which one starts, and is more fully termed I-measurability.)

1. 7. 2. What has already been established concerning the behaviour of I on $\mathscr{L}$ renders it possible to verify the following assertions:

(1) The measurable sets form a σ-algebra which contains all open sets, and therefore all Borel sets too.

(2) The set-function μ is non-negative (and therefore has BV) on the σ-algebra of measurable sets, and μ is there σ-additive. Thus μ (or, rather, its restriction to the set of Borel sets) is a Borel measure.

1. 7. 3. Exercise. Verify the above statements.
[Hints: Notice that $c_{A \cup B} = \inf(c_A + c_B, \underline{1})$, that $c_{A \cup B} = c_A + c_B$ if A and B are disjoint, and that $c_{X \setminus A} = \underline{1} - c_A$. So check that the set $\mathscr{M}$ of measurable sets is an algebra and that μ is additive on $\mathscr{M}$. For the rest notice that

$$c_{\left(\bigcup_{n=1}^{\infty} A_n\right)} = \lim_m c_{\left(\bigcup_{n=1}^{m} A_n\right)}$$

and use Theorem 1. 6. 6.]

1. 7. 4. To these statements may be added a third, namely:

(3) μ is regular (see 1. 2. 1).

To prove this, it suffices (since μ is non-negative) to show that if A is a measurable subset of X and if $\varepsilon > 0$, an open set $B \supseteq A$ exists for which $\mu(B) < \mu(A) + \varepsilon$. Now c_A is integrable and so there exists for any $\varepsilon' > 0$ a function $\phi \in \Phi$ such that $\phi \geq c_A$ and $I(\phi) \leq I(c_A) + \varepsilon'$. It may be assumed that $\phi \leq \underline{1}$ (otherwise replace ϕ by $\inf(\phi, \underline{1})$). The set $B = \{x \in X : \phi(x) > 1 - \varepsilon'\}$ is then open and contains A, and $c_B \leq (1 - \varepsilon')^{-1}\phi$. Hence

$$\begin{aligned}\mu(B) = I(c_B) &\leq (1 - \varepsilon')^{-1} I(\phi) \leq (1 - \varepsilon')^{-1}[I(c_A) + \varepsilon'] \\ &= (1 - \varepsilon')^{-1}[\mu(A) + \varepsilon'] \\ &\leq \mu(A) + \varepsilon'(1 - \varepsilon')^{-1}[1 + \mu(X)],\end{aligned}$$

which in turn is majorised by $\mu(A) + \varepsilon$ provided ε' is chosen small enough (depending on ε and μ). Regularity is thereby established.

1.7.5. To complete the proof of the RRT, it remains only to show that $I(f) = \int f d\mu$ for $f \in C$. In doing this, it may be assumed that $f \geq \underline{0}$. Suppose the range of f is contained in the bounded half-open interval $J = [0, c)$. Partition J into a finite number of similar such intervals, say the intervals J_k. For each k, let $s_k = \inf J_k$. Each set $f^{-1}(J_k)$ is measurable, being the intersection of a closed set with an open set. The function $\sum_k s_k \cdot c_{A_k}$, where $A_k = f^{-1}(J_k)$, is integrable and a member of $B_{\mathcal{M}}(X)$. Taking now a sequence of such partitions of J, each finer than its predecessor, and such that also the maximum length of the sub-intervals converges to zero, one obtains a sequence (f_n), wherein each f_n is non-negative, integrable and belongs to $B_{\mathcal{M}}(X)$, and such that $f_n \uparrow f$ and $f_n \to f$ uniformly on X.

For each n, $I(f_n) = \int f_n d\mu$ as a consequence of the way in which μ is defined in terms of I. As $n \to \infty$, $\int f_n d\mu \to \int f d\mu$ (see Exercise 1.1.8). At the same time, $I(f_n) \to I(f)$, as one sees by appeal to Theorem 1.6.6 (for example). Thus $I(f) = \int f d\mu$, and the proof is complete.

1.7.6. **Exercise.** Prove that:

(1) a subset E of X is negligible if and only if E is measurable and $\mu(E) = 0$;

(2) a countable union of negligible sets is negligible;

(3) an arbitrary union of open, negligible sets is negligible.

As a consequence of (3), there exists a largest open set of measure zero: its complement is termed the support of I (or of μ).

Take $X = [0, 1]$ with its usual topology and

$$I(f) = \int_0^1 f(x)dx ,$$

the Riemann integral, for $f \in C(X)$. What is the support of I? What if $X = [0, 1]$ and

$$I(f) = f(0) + \tfrac{1}{2}f(1) + \int_{1/4}^{3/4} f(x)dx \ ?$$

1.7.7. Exercise. Prove that if two non-negative functions f and g (finite-valued or not) satisfy $f \le g$ save perhaps at the points of a negligible set, then $I^{**}(f) \le I^{**}(g)$.

Deduce that if $f, g \in \mathbf{R}^X$, and if $f = g$ save perhaps at the points of a negligible set, then g is integrable if and only if f is integrable, in which case $I(g) = I(f)$.

[Hint: Note that $f \le \lim_n (g + nc_E)$, if $f \le g$ save perhaps at the points of E.]

1.7.8. Exercise. Let $I = \sum_{n=1}^{\infty} c_n \varepsilon_{x_n}$; see Exercise 1.4.4. Put $S_k = \{x_n : 1 \le n \le k\}$ and $S = \{x_n : n \ge 1\}$. Prove the following statements:

(i) $I^{**}(c_{S_k}) = \sum_{n=1}^{k} c_n$;

(ii) $I^{**}(c_S) = \sum_{n=1}^{\infty} c_n$;

(iii) $X \setminus S$ is negligible;

(iv) $\mu(\{x_n\}) = c_n$;

(v) for any non-negative function f on X,

$$I^{**}(f) = \sum_{n=1}^{\infty} c_n f(x_n) ;$$

(vi) $f \in R^X$ is integrable if and only if $\sum_{n=1}^{\infty} c_n |f(x_n)| < \infty$, in which case

$$I(f) = \sum_{n=1}^{\infty} c_n f(x_n) ;$$

(vii) any subset $A \subseteq X$ is measurable, and

$$\mu(A) = \sum_{n \in A'} c_n ,$$

where $A' = \{n : x_n \in A\}$.

1. 7. 9. Concerning terminology. Having established the existence part of 1. 2. 3 in detail, we may take the existence part of 1. 2. 2 as proven. For reasons which will appear in 1. 10. 4 below, some writers (especially the French) attach the name Radon measure on X to any CLF F on C(X), and to the corresponding σ-measure μ. (The context usually makes it plain whether the functional or the measure-function is in question.) We will follow this practice and denote by M(X) the linear space of all Radon measures on X. It is important precisely because it is the dual of C(X).

In addition, we shall sometimes use the term 'integral' to mean 'non-negative Radon measure'.

In relation to a given Radon measure μ on X, it is standard to say that a property of points of X holds μ - almost everywhere (a. e., for short) if and only if the set of points of X which do not possess the said property is negligible (in the sense explained in 1. 6. 7 above). For example, two functions are said to agree a. e. if and only if the set of points at which they disagree is negligible (i. e., if and only if their difference is a negligible function; see 1. 6. 7 above). Again, any extended real-valued function which is integrable is real- (i. e., finite-) valued a. e.; and a function is negligible if and only if it is zero a. e.

1. 7. 10. Measurable functions. A real-valued function f on X is said to be measurable if and only if, for every $r \in R$, the set $\{x \in X : f(x) > r\}$ is measurable. (The same criterion is applied to extended real-valued functions on X.) As in the case of sets (see 1. 7. 1) this concept of measurability depends on the particular non-negative Radon measure I (or μ) from which one starts. It can happen (cf. Exercise 1. 7. 8) that all functions are measurable; in general, this is not the case. (However, it is in every case true that the production of nonmeasurable sets or functions involves a highly nonconstructive procedure; see Halmos [1], pp. 69-70; Exercise 2. 2. 14 below and Edwards [3], Exercise 3. 19.)

The routine properties of measurable functions (which in the main derive from analogous properties of measurable sets... see 1. 7. 2-1. 7. 4 above) may be found in any text on Lebesgue-type measure and integration theory. In summary, the set of measurable functions is closed under the elementary algebraic operations (insofar as these lead to functions at all) and under all the normal countable limiting processes (taking sups, infs, lim sup and lim inf of sequences, for example); see also Exercise 1. 7. 12. Every continuous and every semicontinuous function is measurable; cf. the argument in 1. 7. 5 above.

Concerning the connections between measurability and integrability of functions (with respect to measures and integrals of the type considered here), the principal results are as follows:

(i) every integrable function is measurable;

(ii) a bounded function is integrable if and only if it is measurable (cf. 1. 6. 4);

(iii) $f \in \mathscr{L}^p$ $(1 \le p < \infty)$ if and only if f is measurable and either $|f|^p \in \mathscr{L}^1$ or $N_p(f) < \infty$.

The proofs are not trivial.

1. 7. 11. Exercise. Let (f_n) be a sequence of real-valued measurable functions on X, and let E be the set of points $x \in X$ for which the sequence $(f_n(x))$ is convergent in **R**. Show that E is measurable.

1. 7. 12. Exercise. Suppose that $f: X \to \mathbf{R}$ is measurable and that $g: \mathbf{R} \to \mathbf{R}$ is continuous. Prove that $g \circ f$ is measurable.

Remark. A little surprisingly, perhaps, $f \circ g$ need not be measurable, see Halmos [1], p. 83.

1.7.13. Exercise. Let $p \in [1, \infty]$ and let $(m, n) \in \mathbf{N} \times \mathbf{N} \mapsto a_{m,n}$ be a double sequence of positive numbers such that

$$\lim_{m\to\infty,\ n\to\infty} a_{m,n} = 0 .$$

Select any strictly increasing sequence $k \mapsto n_k$ of positive integers such that

$$\sum_{k=1}^{\infty} (k^2 a_{n_k, n_{k+1}})^p < \infty .$$

Let $(f_n)_{n \in N}$ be any sequence of elements of $\mathscr{L}^p$ such that

$$N_p(f_m - f_n) \leq a_{m,n}$$

for every $m, n \in N$. Show that the subsequence $(f_{n_k})_{k=1}^{\infty}$ has the property that, for some negligible set $E \subseteq X$, $\lim_{k\to\infty} f_{n_k}(x)$ exists in $\mathbf{R}$ for every $x \in X \setminus E$. Show also that, if f is any real-valued function on X such that $f(x) = \lim_{k\to\infty} f_{n_k}(x)$ for almost every $x \in X \setminus E$, then $f \in \mathscr{L}^p$ and

$$\lim_{n\to\infty} N_p(f - f_n) = 0 .$$

Deduce that $\mathscr{L}^p$ is complete for the seminorm N_p.

[Hints: Consider the sets

$$E'_k = \{x \in X : |f_{n_k}(x) - f_{n_{k+1}}(x)| \geq k^{-2}\} ,$$

$$E_k = E'_k \cup E_{k+1} \cup \ldots, \quad E = \bigcap_{k=1}^{\infty} E_k .$$

Use the properties cited in 1.7.10 and Fatou's lemma 1.5.4.]

1.7.14. Exercise. Suppose that X, I and μ are such that μ is continuous (= diffuse in Bourbaki's language), i.e., $\mu(\{x\}) = 0$ for every $x \in X$. Show how to construct a sequence $(f_n)_{n \in N}$ of non-negative continuous functions on X such that

(i) $\lim_{n\to\infty} I(f_n^p) = 0$ for every $p \in (0, \infty)$

and

(ii) the sequence $(f_n(x))_{n \in N}$ is convergent for no $x \in X$.

Compare this with Exercise 1. 7. 13 immediately above.

Remark. The hypotheses are fulfilled whenever X is a compact Hausdorff topological group G and I is the normalised Haar integral on G (see 2. 1 below). In this case it follows from 2. 9. 6 below that the f_n can be taken to be non-negative trigonometric polynomials on G.

1. 8. Lebesgue's convergence theorem

The discussion in 1. 9 relies in part on establishing what is perhaps the central convergence theorem for Lebesgue-type integrals. This result is now within easy reach.

1. 8. 1. Theorem. Suppose that (f_n) is a sequence of integrable functions which is 'dominated', in the sense that an integrable function g exists for which $|f_n| \leq g$ almost everywhere for each n. Then $\limsup_n f_n$ and $\liminf f_n$ are integrable, and

$$I(\liminf f_n) \leq \liminf I(f_n) \leq \limsup I(f_n) \leq I(\limsup f_n) .$$

In particular, if $f = \lim f_n$ exists, then f is integrable and

$$I(f) = \lim I(f_n) .$$

Proof. By Exercise 1. 7. 7, we may and will suppose that $|f_n| \leq g$ everywhere for each n. Putting $F_m = \sup\{f_n : n \geq m\}$ and

$$F_{mp} = \sup\{f_n : m+p \geq n \geq m\} ,$$

one has $F_{mp} \uparrow F_m$ as $p \to \infty$. Furthermore, $-g \leq F_{mp} \leq g$.

On combining Exercise 1. 6. 5 and the remark following Theorem 1. 6. 6, it appears that F_m is integrable. On the other hand, $F_m \downarrow \limsup f_n$ as $m \to \infty$, and $-g \leq F_m \leq g$. Using again the remark following Theorem 1. 6. 6, this time applied to the functions $-F_m$, it appears that $\limsup f_n$ is integrable. Since $\liminf f_n = -\limsup(-f_n)$, it too is integrable.

The relation

$$I(\liminf f_n) \leq \liminf I(f_n)$$

now appears by applying Exercises 1. 5. 4 and 1. 6. 3 to the functions $f_n + g \geq \underline{0}$ and using linearity of I. Replacing f_n by $-f_n$, it follows that

$$I(-\limsup f_n) \leq \liminf I(-f_n) ,$$

i. e.,

$$I(\limsup f_n) \geq \limsup I(f_n) .$$

The chain of inequalities stated in the theorem is thereby established.

1. 9. Concerning the necessity of the hypotheses in the RRT

There are two questions which are now almost unavoidable, namely

What happens if X is compact but not Hausdorff?
What happens if X is non-compact?

These will be discussed in turn. It will appear that the Hausdorff axiom is scarcely worth regarding as essential, but that compactness is nearly so.

1. 9. 1. X compact non-Hausdorff. Suppose that X is compact but not Hausdorff. It will then in general fail to be c. r. (= completely regular), so that no longer will it be necessarily true that every non-negative lower semicontinuous function on X is the upper envelope of functions in C(X). In fact, there exist infinite compact spaces X on which the only continuous functions are constants, and yet on which there exist non-constant non-negative lower semicontinuous functions, and of which <u>all</u> subsets are Borel sets. As a consequence of this, the concept of Borel measure is not always ideally suited to the statement of the RRT.

Nonetheless, one may <u>define</u> Φ to consist of just those non-negative functions on X which are upper envelopes of subsets of C(X) and then repeat 1. 4, 1. 5 and 1. 6 verbatim.

As in 1.7.2, the set $\mathscr{M}$ of all measurable subsets of X is still a σ-algebra. It may fail to contain all Borel subsets of X. However, in order that a σ-measure μ on $\mathscr{M}$ be suitable for the representation of linear functionals on C(X), it is enough that $C(X) \subseteq B_{\mathscr{M}}(X)$. Now the smallest σ-algebra $\mathscr{Z}$ with the property that $C(X) \subseteq B_{\mathscr{Z}}(X)$ can be shown (see Exercise 1.9.10) to be that generated by the zero-sets $f^{-1}(\{0\})$ of functions $f \in C(X)$. It may be verified (see Exercise 1.9.11) that in all cases $\mathscr{M}$ contains $\mathscr{Z}$, so that an integral representation is again possible. If X is c.r., $\mathscr{Z}$ is identical with the σ-algebra of so-called Baire sets, i.e., the σ-algebra generated by the compact G_δ-sets (see Halmos [1], pp. 216 et seq.).

Lebesgue's theorem 1.8.1 demands no modification whatsoever.

1.9.2. It is some comfort to recognise that the RRT can be reformulated in a way which avoids the rather bewildering complexity of diverse σ-algebras and σ-measures upon them, and this without losing any of the essential value of the theorem.

Given a NNLF I on C(X), a procedure has been laid out which leads to a non-negative linear extension of I having as domain an entity $\mathscr{L}$ with the following properties:

(i) $\mathscr{L}$ is a linear subspace of $\mathbf{R}^X$ which contains C(X);

(ii) if a sequence (f_n) of elements of $\mathscr{L}$ converges boundedly to $f \in \mathbf{R}^X$, then $f \in \mathscr{L}$;

(iii) if the sequence (f_n) is as in (ii), then

$$\lim I(f_n) = I(f) .$$

The linear space $\mathscr{L}$ depends upon I.

Amongst those linear subspaces L of B(X) which contain C(X), and which are stable under the formation of limits of boundedly convergent sequences of members of L, there is a smallest. Denote this subspace of B(X) by L(X). Then $\mathscr{L} \supseteq L(X)$, and the extension of I enjoys the property

(P) If $f_n \in L(X)$ $(n = 1, 2, \ldots)$, and if the sequence (f_n) converges boundedly to a function $f \in \mathbf{R}^X$, then $f \in L(X)$ and $\lim I(f_n) = I(f)$.

It is easily verified that, for a given NNLF I on C(X), there exists only one linear extension of I to L(X) for which (P) is true. It has been seen that this unique extension is non-negative on L(X).

The promised reformulation of the RRT is as follows:

1.9.3. **Theorem.** Let X be a compact space (Hausdorff or not), and let I be a NNLF on C(X). Then I admits a unique linear extension to L(X) which satisfies (P); this extension is non-negative on L(X).

1.9.4. **Exercise.** Prove the statement made above concerning the uniqueness of the linear extension of I to L(X), the extension being assumed to satisfy (P).
[Hint: Suppose I' and I" are extensions of the stated type. Consider the set L of $f \in L(X)$ for which $I'(f) = I''(f)$. Verify that L is a linear subspace of B(X) which contains C(X) and which is stable under the formation of limits of boundedly convergent sequences of its members.] □

By way of preface to the next corollary it is necessary to recall that a sequence (f_n) extracted from a normed linear space H is defined to be weakly convergent in H to an element f of H, if and only if

$$\lim F(f_n) = F(f)$$

for each CLF F on H.

1.9.5. **Corollary.** Let X be a compact space. In order that a sequence (f_n) extracted from C(X) be weakly convergent in C(X) to $f \in C(X)$, it is sufficient that $\lim f_n = f$ boundedly on X.

Proof. Inasmuch as each CLF on C(X) is expressible as the difference of two NNLFs, this follows directly from Theorem 1.9.3.

1.9.6. **Remark.** The stated conditions in 1.9.5 are also necessary. That the pointwise convergence of (f_n) to f is necessary, follows from the fact that, for each $x \in X$, $f \mapsto f(x)$ is a NNLF on C(X). The necessity of the boundedness restriction (i.e., the demand that $\sup_n \|f_n\| < \infty$), is a special case of the Banach-Steinhaus theorem (Edwards [2], 7.1.1 or 7.1.3). It will not be proved here.

1.9.7. X non-compact. It will now be shown that there is a much more genuine breakdown of the RRT when the hypothesis of compactness is relaxed.

1.9.8. Theorem. Suppose X is a topological space satisfying the following condition:

(i) There exists a sequence (x_n) of points of X, and a sequence (f_n) of elements of $C(X)$, such that $\lim f_n = \underline{1}$ boundedly on X and $\lim_{m\to\infty} f_n(x_m) = 0$ for each n.

Then there exists on $C(X)$ a NNLF I which admits no linear extension to $L(X)$ satisfying condition (P).

The condition (i) implies that X is non-compact. On the other hand, (i) is satisfied whenever X is non-compact, c.r., and expressible as the union of a sequence (W_n) of relatively compact, open subsets of X.

Proof. Assuming that (i) is fulfilled, define for $f \in C(X)$ the number

$$p(f) = \limsup_{n\to\infty} f(x_n) .$$

Evidently,

$$p(f + g) \le p(f) + p(g) ,$$
$$p(cf) = c.\, p(f) ,$$

if $f, g \in C(X)$ and c is a non-negative real number. According to the Hahn-Banach theorem (Exwards [2], 1.7.1), there exists a linear functional I on $C(X)$ such that

$$I(f) \le p(f) .$$

This is easily seen to entail that

$$\liminf_n f(x_n) \le I(f) \le \limsup_n f(x_n) \qquad (1.9.1)$$

for each $f \in C(X)$; in particular, I is a NNLF on $C(X)$.

Suppose that this I were extendable to $L(X)$, subject to (P). On the one hand, from (1.9.1), $I(\underline{1}) = 1$. From (P), $I(\underline{1}) = \lim_m I(f_m)$. From

(1.9.1) again, $I(f_m) = 0$ for each m. Thus there appears the absurdity:

$$1 = I(\underline{1}) = \lim_m I(f_m) = \lim_m 0 = 0 .$$

The conclusion is that I is indeed not extendable. This completes the proof of the first assertion.

It will next be shown that (i) entails that X is non-compact. As before, one infers the existence of a linear functional I on $C(X)$ satisfying (1.9.1). This I is evidently positive and continuous. As has been seen, $I(\underline{1}) = 1$ and $I(f_n) = 1$ for each n. Were X compact, Corollary 1.9.5 would show that $\lim f_n = \underline{1}$ weakly in $C(X)$, and therefore that $1 = I(\underline{1})$ would equal $\lim I(f_n)$, which is 0. This contradiction forces the conclusion that X is non-compact.

Finally assume that X is non-compact, c.r., and expressible as the union of a sequence (W_n) of relatively compact open sets. It may be assumed that the W_n increase with n. Moreover, by dropping terms of the sequence (W_n), if necessary, it may be arranged that $\overline{W}_n$ is a proper subset of W_{n+1}. Choose arbitrarily x_n from $W_{n+1} \backslash \overline{W}_n$. On the other hand, since X is c.r., there exists for each n a continuous function f_n from X into $[0, 1]$ such that

$$f_n(\overline{W}_n) = \{1\}, \quad f_n(X \backslash W_{n+1}) = \{0\} ;$$

for this, see Kelley [1], p. 142. It is plain that the sequences (x_n) and (f_n) satisfy the demands imposed by (i).

1.9.9. Remark. The problem of characterising completely those c.r. spaces X for which each NNLF on $C(X)$ is representable by integration with respect to a σ-measure, was solved by Glicksberg [1]. He showed that it is necessary and sufficient for this that X satisfy any one of the following four (equivalent) conditions:

(1) If $f_n \in C(X)$ and $f_n \uparrow f \in C(X)$, then $f_n \to f$ uniformly on X (cf. Dini's theorem 1.3.7).

(2) Each continuous function on X is bounded.

(3) Each continuous function on X assumes a maximum.

(4) Each bounded equicontinuous subset of $C(X)$ is relatively compact therein.

For other work on the representation of linear functionals by integrals, see Hewitt [1], [3].

1.9.10. Exercise. Suppose that X is as in 1.9.1. Verify that the smallest σ-algebra $\mathcal{Z}$ of subsets of X such that $C(X) \subseteq B_{\mathcal{Z}}(X)$ is the σ-algebra $\mathcal{Z}_0$ generated by

$$\{f^{-1}(\{0\}) : f \in C(X)\} .$$

1.9.11. Exercise. Suppose that X, Φ and $\mathcal{M}$ are as in 1.9.1, that $f \in C(X)$ and that $E = f^{-1}(\{0\})$. Show that $c_E = \underline{1} - \phi$ for some $\phi \in \Phi$, and deduce that $E \in \mathcal{M}$; cf. 1.6.4 and 1.7.1.

1.10. Historical remarks

The first of the C(X) spaces to receive attention was, as would be expected, that in which X is a compact interval on the real axis. The interval [0, 1] is typical, the corresponding space being denoted by C[0, 1], rather than C([0, 1]).

1.10.1. It would appear that Hadamard ([1], [2]) was the first to consider the representation of CLFs on C[0, 1], though this particular problem was only a small part of his concern. A resume of some of this early work is to be found in Volterra [1].

Hadamard's result asserts that, if F is a CLF on C[0, 1], then there exists a sequence (K_n) extracted from C[0, 1] such that

$$F(f) = \lim_n \int_0^1 fK_n , \qquad (1.10.1)$$

each integral being a Riemann integral. This result appeared in 1903. Fréchet ([1], [2], [3], covering the years 1904-1907) re-proved the result, noted certain restrictions which may be imposed on the K_n, and remarked further that F can be represented in a different way, namely,

$$F(f) = \lim_n \sum_{r=1}^{n} k_{nr} f(r/n) \qquad (1.10.2)$$

for a suitably chosen double sequence (k_{nr}). Both Hadamard and Fréchet gave applications of these representation formulae.

1.10.2. Both formulae (1.10.1) and (1.10.2) exhibit the defect of non-uniqueness. Perhaps the most satisfying way of seeing this is to consider briefly a proof of (1.10.1). Let Q denote the square $[0, 1] \times [0, 1]$ and take a sequence (u_n) of non-negative continuous functions on Q such that

$$\int_0^1 u_n(x, y)dy = 1 \qquad (0 \leq x \leq 1) ,$$

and such that, for each $t > 0$,

$$\lim_n u_n (x, y) = 0$$

uniformly for $(x, y) \in Q$ and $|x-y| \geq t$. As examples one might take

$$u_n(x, y) = \sup (0, n-n^2|x-y|) ,$$

or

$$u_n(x, y) = e^{-n^2(x-y)^2} / \int_0^1 e^{-n^2(x-t)^2} dt .$$

Routine analysis then confirms that, if $f \in C[0, 1]$, and if the sequence (f_n) is defined by

$$f_n(x) = \int_0^1 f(y)u_n(x, y)dy ,$$

then (f_n) converges in $C[0, 1]$ to f, as $n \to \infty$. On the other hand, by uniform continuity of u_n on Q, f_n is the limit in $C[0, 1]$ of sums

$$f_{np}(x) = \sum_{r=1}^{p} f(r/p)u_n(x, r/p). 1/p .$$

Let F be a CLF on $C[0, 1]$. For $0 \leq y \leq 1$, denote by $U_n(y)$ that element of $C[0, 1]$ which is the function $x \mapsto u_n(x, y)$. Uniform continuity of u_n entails that $y \mapsto U_n(y)$ is continuous from $[0, 1]$ into $C[0, 1]$. Hence

$$K_n : y \mapsto F(U_n(y))$$

is a member of C[0, 1]. Now linearity of F yields

$$F(f_{np}) = \sum_{r=1}^{p} f(r/p)K_n(r/p). 1/p .$$

As $p \to \infty$, the left-hand member of this equality converges to $F(f_n)$ (since F is continuous); the right-hand member converges to $\int_0^1 fK_n$. As $n \to \infty$, $F(f_n) \to F(f)$, and (1.10.1) ensues.

Notice that, when F is given, the sequence (K_n) can be chosen in many ways inasmuch as many sequences (u_n) are available. Whence the non-uniqueness referred to above. The u_n may be chosen to be polynomials, in which case the K_n are polynomials. If F is non-negative, the K_n are non-negative too.

Fréchet considered in a rather similar way functionals on C[0, 1] which are not linear, but which are analytic in a suitable sense. These ideas were neglected for quite a while thereafter but were ultimately developed considerably.

1.10.3. Riesz's contributions to the representation theorem began in 1909 with his paper [1]. Here appears the first representation formula of the type

$$F(f) = \int_0^1 f dm , \qquad (1.10.3)$$

the right-hand member of (1.10.3) being a Riemann-Stieltjes integral with respect to the function m having bounded variation on [0, 1]. An easily-accessible proof of this formula is found in Banach [1], pp. 59-61. Since that time, Riesz has given several other proofs of (1.10.3), his papers [2], [3] and [4] covering the period 1911-52. Helly [1] also contributed a proof of (1.10.3).

Until the theory of Riemann-Stieltjes integrals has been developed to a considerable extent (which is certainly no easier than studying the corresponding Lebesgue-Stieltjes integrals), result (1.10.3) is on a par with (1.1.4) with μ a (finitely additive) measure. Moreover, a formulation involving Riemann-Stieltjes integrals is not convenient for extensions from [0, 1] to more general spaces.

1.10.4. In 1913 Radon [1] extended Riesz's formula from [0, 1] to a general compact subset of $\mathbf{R}^n$ and, at the same time, introduced Lebesgue-type integrals with respect to Borel measures. Banach, in Note II to Saks [1] (published in 1937), and Saks himself in the following year (Saks [2]), carried Radon's version over to a general compact metric space. Contemporary with this was Markov's paper [1], which examined integral representations in terms of finitely additive measures, the space X being assumed to be normal but not necessarily Hausdorff.

1.10.5. The version of the RRT appearing as Theorems 1.2.2 and 1.2.3 arrived in 1941 in Kakutani's paper [1]. It would therefore be more accurate to speak of this as the Riesz-Radon-Banach-Saks-Kakutani theorem! A novel approach to this version of the theorem is due to Varadarajan [1].

1.10.6. In all these developments the customary approach to Lebesgue-type integrals had been to the fore: by this is meant the viewpoint which derives an integral from a given measure. In the meantime, however, a novel approach to integration theory had been suggested in 1918 by Daniell [1]. In brief, his idea was to treat an integral as a type of linear functional. A consequence of this would be a formulation of the RRT rather close to that suggested in 1.9.2 and 1.9.3. Daniell's work was largely neglected for some twenty years, at the end of which period interest was revived by Bourbaki (c. 1937), presumably when plans were being laid for the relevant sections of his future work [1] in this field. Publication of this work began in 1952, since when further volumes have appeared. Between 1937 and 1952, some of the ideas filtered through to the mathematical world at large (in the guise mainly of books and research papers authored by individual members of the Bourbaki group, such as Weil and H. Cartan). Other writers, notably Stone [1], [2], [3], [4], also found inspiration in Daniell's ideas. As a result, the method was discussed in various papers and books from 1950 onwards: Loomis [1]; Naimark [1]; Hewitt [2]; Edwards [1] and [2], Chapter 4, to mention a few. Bourbaki introduced the term 'Radon measure on X' to describe a CLF on C(X), X being a Hausdorff compact space (see 1.7.9 above); the term 'pre-integral' is perhaps more suggestive (but not standard).

It is characteristic of the work issuing from Daniell's approach that a fully-fledged integration theory (and not merely a study of the RRT) is the principal aim; and that the concept of measure is subordinated to that of integral, a reversal of the traditional relationship. Of course, the RRT can be made to appear as a by-product of the conventional approach, as in Halmos [1]. The Daniell-based approach to the RRT amounts to a constructive extension of the given CLF F from C(X) to a larger space of functions, as is witnessed by the substance of 1.9.

1.11. Complex-valued functions

We shall leave to the reader the task (mainly a routine one) of modifying the preceding material so as to cover complex-valued (rather than real-valued) functions on X.

In the sequel we shall use C(X) to denote the complex-linear space of continuous complex-valued functions on X, C(X) being equipped with the norm $\|f\| = \sup_{x \in X} |f(x)|$. What was hitherto denoted by C(X) will now be denoted by $C_{\mathbf{R}}(X)$, and a similar notational change will be made in relation to the other function-spaces involved ($\mathscr{F}$, $\mathscr{F}^p$, $\mathscr{L}^p$ and L^p). Thus $C_{\mathbf{R}}(X)$ is a real-linear subspace of C(X).

As to integrals and Radon measures, every element of what has hitherto been denoted by M(X), henceforth to be denoted by $M_{\mathbf{R}}(X)$, can be extended uniquely and in an obvious way into a continuous complex-linear and complex-valued functional on C(X); this extension will be denoted by the same symbol. These are the real Radon measures on X. The symbol M(X) will henceforth denote the set of all continuous complex-valued complex-linear functionals on C(X), each such functional being termed a (complex) Radon measure on X.

The expected connections hold between the pairs $\mathscr{F}$ and $\mathscr{F}_{\mathbf{R}}$, $\mathscr{F}^p$ and $\mathscr{F}^p_{\mathbf{R}}$, $\mathscr{L}^p$ and $\mathscr{L}^p_{\mathbf{R}}$, L^p and $L^p_{\mathbf{R}}$, et cetera.

1.11.1. Exercise. Let λ denote a non-negative Radon measure on X, i.e., real-valued real-linear functional on $C_{\mathbf{R}}(X)$ such that $\lambda(f) \geq 0$ for every $f \in C_{\mathbf{R}}(X)$ satisfying $f \geq \underline{0}$. Suppose λ to be extended into a complex-valued complex-linear functional on C(X). Show that

$$|\lambda(f)| \leq \lambda(\underline{1}) \|f\|$$

for every $f \in C(X)$.

Part 2 · Harmonic Analysis on Compact Groups

2.0. Introduction to Part 2

2.0.1. Harmonic analysis might be said to comprise the study of functions and function spaces defined over a topological group G, special reference being paid to the functional operators of translation arising from the group structure of G; see the discussion in Edwards [3], Chapter 2. This description is correct as far as it goes, but it is unlikely to convey much except to those who are already acquainted with the subject (and who therefore have no great need of a description anyway). The only effective way to discover what harmonic analysis is about, is to dip into it, taking stock of just enough (but not too much) detail. Our treatment attempts to present in such a style something from which the taster may choose.

As seems entirely natural, the display offered refers to one of the technically simpler cases: that in which the underlying group G is compact and Hausdorff. (The case in which G is also Abelian is even simpler, and some readers may wish to concentrate on this situation, which still offers many challenging problems.)

2.0.2. Even with the restrictions mentioned in 2.0.1, there remain a number of approaches to abstract harmonic analysis. Their relative merits depend in part on how much is assumed about the underlying group G (always assumed locally compact and Hausdorff), though there are no sharp dividing lines. For compact groups, the various approaches are much on a par with each other and the choice is largely a matter of taste. We mention a few of the possible approaches.

(i) For compact G there is the traditional approach based from the outset on finite dimensional representations of G (suggested by the classical algebraic study of finite groups). This is broadly the method used in Pontryagin [1], Weil [1] and Hewitt and Ross [1] (though the last reference features other methods as well).

(ii) For compact G there is an approach based upon the study of H*-algebras (a special type of non-commutative Banach algebra). This method is featured in Loomis [1], §§27. 39. 40 and Naimark [1], pp. 330, 431 ff.

(iii) For locally compact Abelian groups there is an approach based on the Gelfand theory of commutative Banach algebras. See Hewitt and Ross [1]; Loomis [1], Chapters VI and VII; Naimark [1], §31; Katznelson [1], Chapter VIII; Bourbaki [3].

(iv) For locally compact groups there is an approach due to Godement, Gelfand and Raikov which swings back to representations as a key tool, though now one has to handle infinite dimensional ones. Mixed in with this in an essential way is the study of positive definite functions. Some aspects of this approach are covered in Hewitt and Ross [1], §§21, 22 and Naimark [1], §30.

A little of most of these treatments intrudes into the account to follow, though in the main the approach is that described in (i).

Readers who are completely new to the subject may find that portions of Edwards [3] provide a useful bridge.

In view of Part 1, our natural lead-in is to begin by settling the existence and essential uniqueness of an invariant integral on every compact Hausdorff group. The subsequent development of harmonic analysis makes constant and essential use of this distinguished integration process; see again Chapter 2 of Edwards [3].

2. 0. 3. Prerequisites. It is assumed that the reader has a basic acquaintance with set-theoretic topology and group theory (very few properties of any depth in either discipline will be explicitly used). Granted this, we recall that a topological group is defined (perhaps a little loosely) as a group G and a topology t on G such that the function $(x, y) \mapsto xy^{-1}$ is continuous from $G \times G$ (with the product topology) into G. One usually (and slightly improperly) refers to the 'topological group G', omitting explicit reference to t. Some examples appear in 2. 0. 4 below. Only a few very basic properties of topological groups will be needed for a general understanding of what follows; acquaintance with Hewitt and Ross [1], (4.1)-(4.9) is adequate to sanction a start to be made. (References to other portions of Hewitt and Ross [1] will be made very frequently as we go along; for most things covered in Part 2 of these notes it is, indeed, the standard reference.)

A standing convention from here on is to the effect that (unless the contrary is explicitly mentioned) all topological groups are assumed to be Hausdorff.

2.0.4. Some examples

(i) Any group in the purely algebraic sense can be regarded as a topological group by endowing it with its discrete topology (in which all subsets are open). The resulting topological group is compact if and only if the group is finite.

(ii) The circle group $\mathbf{T}$: this is the multiplicative group of unimodular complex numbers taken with its usual topology (induced on it as a subspace of the complex plane).

$\mathbf{T}$ is isomorphic as a topological group to, and is often identified with, the quotient $\mathbf{R}/2\pi\mathbf{Z}$, where $\mathbf{R}$ is the additive group of real numbers with its usual topology and $\mathbf{Z}$ is the discrete additive group of rational integers; the algebraic quotient group $\mathbf{R}/2\pi\mathbf{Z}$ is taken with the quotient topology. The mapping $t + 2\pi Z \mapsto e^{it}$ is a topological group isomorphism of $\mathbf{R}/2\pi\mathbf{Z}$ onto $\mathbf{T}$. Functions on $\mathbf{T}$ become identified with functions on $\mathbf{R}$ which have period 2π.

Although $\mathbf{T}$ and $\mathbf{R}$ are the most familiar underlying groups for harmonic analysis, they are not entirely typical. In the first place, they are Abelian. In the second place, even among Abelian groups, they are not in all respects fully typical (or even the simplest); cf. 2.9.7 below.

(iii) Among the most important and most familiar non-Abelian groups are the various linear groups, obtained in the following way. Let V be any finite-dimensional real or complex linear space. (Algebraists often start with a linear space over an arbitrary field, but there will be no call for this degree of generality in these notes.) The set End(V) of all endomorphisms of V is finite-dimensional algebra over the same field. The subset GL(V) of End (V), comprising all invertible endomorphisms of V, forms a group under the operation of multiplication (i. e. composition) of endomorphisms. GL(V) is usually termed the general (or full) linear group associated with V.

Choose any norm $\|.\|$ on V and define the usual operator norm of $T \in End(V)$ by

$$\|T\| = \sup \{ \|Ty\| : y \in V, \quad \|y\| \leq 1 \} . \qquad (2.0.1)$$

Use this norm to define a topology on End(V). (The apparent dependence of this topology upon the choice of the norm on V is illusory; see Edwards [5], 1. 1 and 1. 2.) With the induced topology, GL(V) becomes a locally compact group.

Essentially, GL(V) depends only on the field F (= **R** or **C**) over which V is regarded as a linear space and the corresponding dimension n of V. So, one often writes GL(n, F) in place of GL(V). On choosing a base for V, GL(n, F) is frequently realised as the group of invertible $n \times n$ matrices with entries in F, the topology being simply that of convergence (in F) of the entries.

There are numerous important subgroups of GL(V); the following are examples (Hewitt and Ross [1], (2. 7) and (4. 18)):

SL(V), the special linear group comprising those $T \in GL(V)$ with $\det T = 1$;

if V is a real (resp. complex) Hilbert space, one has O(V) (resp. U(V)), the group of orthogonal (resp. unitary) endomorphisms of V; and

SO(V) (resp. SU(V)), the group of special orthogonal (resp. special unitary) endomorphisms of V.

Each of O(V), SO(V), U(V) and SU(V) is a compact group, non-Abelian if $n = \dim V > 1$.

It is not difficult to show that a subgroup G of GL(V) is compact if and only if (a) G is closed in GL(V), and (b) G is bounded in the sense that

$$\sup \{ \|T\| : T \in G \} < \infty .$$

Representation theory is largely concerned with homomorphisms of given groups into various linear groups.

(iv) If $(G_i)_{i \in I}$ is an arbitrary family of compact groups, the product group $G = \Pi_{i \in I} G_i$ is a compact group when taken with its product topology.

One of the principal consequences of the theory to be described below is that every compact group G is isomorphic to a closed subgroup of a suitable product $\Pi_{i \in I} U(d_i)$, where I is some (generally infinite) index set, each d_i is a positive integer, and $U(d_i)$ is the unitary group (see (iii) above) associated with a complex Hilbert space of dimension d_i.

See also 2.2.13 below.

2.1. Invariant integration

In what follows G will, except in the relatively few places devoted to 'asides' (at which points explicit notice will be given), denote a compact (Hausdorff) group.

2.1.1. On G, as on any compact Hausdorff space, there are many non-zero integrals (i.e., non-zero non-negative linear functionals on C(G) or non-zero non-negative Radon measures on G; cf. 1.2 and 1.7.9 above). The group structure of G induces certain translation operators on C(G), namely, the <u>left translation operators</u> $f \mapsto L_a f$ and the <u>right translation operators</u> $f \mapsto R_a f$, where, for each $a \in G$ and each $f \in C(G)$,

$$L_a f : x \mapsto f(a^{-1}x), \quad R_a f : x \mapsto f(xa^{-1}) . \qquad (2.1.1)$$

Each L_a and each R_a is a continuous endomorphism of C(G) and one may ask whether there exist any non-zero integrals I which are either

(i) <u>left invariant</u> in the sense that $I \circ L_a = I$ for every $a \in G$,

or (ii) <u>right invariant</u> in the sense that $I \circ R_a = I$ for every $a \in G$,

or perhaps both left and right invariant... <u>bi-invariant</u> is the term to be used to describe this third state of affairs. If a non-zero left (or right) invariant integral I exists, it may be <u>normalised</u>, i.e., multiplied by a positive number so chosen as to yield a left (or right) invariant integral I such that $I(\underline{1}) = 1$.

The problem is thus that of the existence of normalised left (or right, or bi-) invariant integrals on G; and, if such exist, the problem of classifying them.

2.1.2. The complete answer to these problems is as follows: for a given compact group G,

(i) there exists precisely one normalised left invariant integral I and every left invariant integral is of the form cI, where c is a non-negative number;

(ii) I is also the unique normalised right invariant integral (hence also the unique normalised bi-invariant integral);

(iii) I is reflection invariant, i.e. $I(\check{f}) = I(f)$ for every $f \in C(G)$, where $\check{f} : x \mapsto f(x^{-1})$;

(iv) $I(f) > 0$ whenever $f \in C_{\mathbf{R}}(G)$ $f \geq \underline{0}$, $f \neq \underline{0}$.

The reader should pause to derive (iii) and (iv) from (i) and (ii).

In establishing (i)-(iv), it is plainly enough to consider the behaviour of I on $C_{\mathbf{R}}(G)$. With this restriction made, a lucid and easy-to-read proof of (i)-(iv) appears in Pontryagin [1], pp. 91-8; see also Exercise 2.1.18. (Although this proof is not in the very latest spirit, it has the advantage of being less sophisticated than others intended to cover the wider fields mentioned in 2.1.3, for example.) This proof is partially constructive, insofar as it produces I(f) for any given $f \in C_{\mathbf{R}}(G)$ as the unique number c such that $\underline{c}$ belongs to the closed convex envelope in C(G) of $\{L_a f : a \in G\}$ (or of $\{R_a f : a \in G\}$). This same characterisation of I(f) remains valid for arbitrary $f \in C(G)$, but no use will be made of it in the sequel.

2.1.3. **Remarks.** (i) Pontryagin's proof mentioned in 2.1.2 suffices to produce the mean value for uniformly almost periodic functions on an arbitrary group (concerning which see especially Maak [1]).

(ii) For locally compact groups G, it is still true that there exist non-zero left invariant (resp. right invariant) integrals (whose initial domain of definition is the space $C_c(G)$ of continuous complex-valued functions f on G for which the support $\{x \in G : f(x) \neq 0\}^-$ is compact). These invariant integrals are often termed <u>Haar integrals</u>, incidentally. It is also still true that any two such non-zero left (resp. right) invariant integrals are each a non-zero constant multiple of the other. However, unless G is compact, there is no natural way of normalising such integrals and so picking out a distinguished one; nor, in general, is a left invariant integral also either right invariant or reflection invariant. For details, see Hewitt and Ross [1], Chapter 4; Weil [1]; Edwards [2], Chapter 4; Bourbaki [2], Chapter 7; to mention only a few of the possible references.

Exercise 2.1.17 indicates what may happen for groups G which are not locally compact.

2.1.4. Concerning notation

(i) In what follows, the σ-measure derived in the manner described in 1.7 above from the normalised invariant integral on a (compact) group G will be denoted by μ or μ_G; μ_G is usually termed the normalised Haar measure on G. The number I(f) is accordingly (cf. 1.7.5 above) often denoted by

$$\int_G f d\mu \quad \text{or} \quad \int_G f d\mu_G ,$$

or by the expressions resulting from these by suppression of explicit reference to G (when the latter is understood from the context). The 'bound variable' expressions

$$\int f(x) d\mu(x) \quad \text{and} \quad \int f(x) dx$$

will also appear frequently.

In terms of μ, the invariance properties of the integral are equivalent to those expressed in the formulae

$$\mu(aE) = \mu(Ea) = \mu(E^{-1}) ,$$

valid for every μ-measurable subset E of G and every $a \in G$; and 2.1.2 (iv) is equivalent to the property that $\mu(U) > 0$ for every non-void open subset U of G, as a result of which $\underline{0}$ is the only continuous function which is negligible.

(ii) Integrability and measurability (without further qualification must be understood to mean these concepts in relation to the normalised invariant integral and the corresponding Haar measure μ. Likewise, $\mathscr{L}^p(G)$ and $L^p(G)$ will denote the Lebesgue spaces formed in relation to the normalised invariant integral; cf. 1.6.7 above.

In accordance with the penultimate paragraph of 1.6.7, we shall almost always fail to make the proper distinction between real- or complex-valued functions on G and the corresponding equivalence classes (modulo equality a. e.). For instance, in the second paragraph of 2.1.5 below, it would be more correct to begin by saying that, if f, g $\in \mathscr{L}^1(G)$ (rather than $L^1(G)$), then the stated procedure leads to a function h defined and equal a. e. to an element of $\mathscr{L}^1(G)$; and to note then that the equivalence class of h depends solely on, and is uniquely determined by, the classes of f and g; at which point one may pass to the quotient space $L^1(G)$, and so on. From now on such licence will be taken without special comment (but it is hoped that the reader will make the mental digression where it is necessary).

(iii) There is one trick which should be mentioned in passing, namely, that of injecting $L^1(G)$ into $M(G)$, so that one may think of integrable functions as special cases of Radon measures. This is done by associating with each $f \in L^1(G)$ the Radon measure $\mu^f = f\mu \in M(G)$ defined by

$$\mu^f : g \mapsto \int_G fg d\mu ;$$

symbolically, $d\mu^f(x) = f(x)d\mu(x)$. This injection is obviously linear, and it is not difficult to show that it is an isometry. The image of $L^1(G)$ is not the whole of $M(G)$, unless G is discrete; the detailed characterisation of elements of the image is relatively complicated and will not be needed in the sequel. (See Hewitt and Ross [1], (12.17); Edwards [2], Section 4.15.)

(iv) At this point account has to be taken of the fact that the translation operators may be regarded as acting on each $L^p(G)$ and $M(G)$ as well as on $C(G)$ (see 2.1.1). In the case of $L^p(G)$, the required definition is made in the manner suggested by (2.1.1), noting that $L_a f$ and $R_a f$ (defined as in (2.1.1) for any complex-valued function f) are negligible whenever f is negligible. How $L_a\mu$ and $R_a\mu$ are to be defined when $\mu \in M(G)$ is suggested by (iii) immediately above, namely:

$$L_a\mu : g \mapsto \mu(L_{a^{-1}}g) ,$$
$$R_a\mu : g \mapsto \mu(R_{a^{-1}}g)$$

for every $g \in C(G)$.

In all cases, the translation operators are isometries (as they were when regarded as endomorphisms of C(G)).

As will be discussed in 2.1.10, a semi-qualitative approach to the behaviour of these function spaces under the action of the translation operators can provide a useful key to what in broad outline harmonic analysis is all about. Meanwhile, a few more technical matters arising from and related to invariant integrals demand attention.

2.1.5. **Convolutions.** This topic may be treated much as in Edwards [3], Chapter 3, though careful attention has to be paid to non-commutativity. More general accounts will be found in Hewitt and Ross [1], §§ 19, 20 and Bourbaki [2], Chapter 8.

If $f, g \in L^1(G)$, it may be shown that for almost every $x \in G$ the function

$$y \mapsto f(y)g(y^{-1}x)$$

is integrable, and that the function defined almost everywhere on G by the formula

$$x \mapsto \int f(y)g(y^{-1}x)dy$$

is in $L^1(G)$; this function is denoted by $f * g$ and is termed the convolution of f and g (in that order):

$$f * g(x) = \int f(y)g(y^{-1}x)dy \quad \text{a. e.} \tag{2.1.2}$$

Moreover,

$$\|f * g\|_1 \leq \|f\|_1 \|g\|_1 \tag{2.1.3}$$

for $f, g \in L^1(G)$.

The mapping $(f, g) \mapsto f * g$ is bilinear from $L^1(G) \times L^1(G)$ into $L^1(G)$; (2.1.3) expresses the continuity of this bilinear map.

Except when G is Abelian, convolution is not commutative. However, it is true that

$$f * g = (\check{g} * \check{f})^{\vee} \tag{2.1.4}$$

for every $f, g \in L^1(G)$; see Exercise 2.1.15.

Convolution is associative: a direct proof is possible, using the Fubini-Tonelli theorem; alternatively, see Exercise 2.4.10.

If $f \in L^p(G)$ and $g \in L^{p'}(G)$, where $1 \le p \le \infty$ and $1/p + 1/p' = 1$ (interpreted so that $1' = \infty$ and $\infty' = 1$), $f * g$ is (equal a. e. to) a continuous function (which is uniquely determined; see the end of 2.1.4(i)), and

$$\|f * g\|_\infty \le \|f\|_p \|g\|_{p'} . \tag{2.1.5}$$

In these circumstances it is usual to agree that $f * g$ denotes the said continuous function.

If both f and g are integrable and at least one is (equal a. e. to a function which is) continuous, $f * g$ is (equal a. e. to a function which is) continuous. Here again $f * g$ is usually understood to denote this continuous function.

If $f \in L^1(G)$ and $g \in L^p(G)$, then $f * g \in L^p(G)$ and

$$\|f * g\|_p \le \|f\|_1 \|g\|_p ; \tag{2.1.6}$$

in view of (2.1.4), there is a similar assertion applying when the order of the convolution factors is reversed.

It thus appears that each of $C(G)$ and $L^p(G)$ is an algebra over the complex field when products are taken to mean convolution: this structure comes to the fore when ideals are discussed in 2.12. [$C(G)$ is also an algebra under pointwise multiplication; the same is not true of $L^p(G)$ when $p \ne \infty$, unless G is discrete; see Exercise 2.1.16.]

There are close connections between translation and convolution. To begin with,

$$f * R_a g = R_a(f * g), \quad L_a f * g = L_a(f * g) \tag{2.1.7}$$

whenever $f, g \in L^1(G)$ and $a \in G$. Also, if $f \in L^1(G)$ and $g \in L^p(G)$ or $C(G)$), $f * g$ is the limit in $L^p(G)$ (or in $C(G)$) of linear combinations of left translates $L_a g$ of g; there is an analogous assertion about $f * g$ and right translates $R_a f$ of f, whenever $f \in L^p(G)$ (or $C(G)$) and $g \in L^1(G)$. See Edwards [3], 3.1.8-3.1.9.

Finally, convolution may be extended so as to appear as a bilinear mapping of $M(G) \times M(G)$ into $M(G)$; space precludes further discussion here, but see e.g. Edwards [2], Section 4.19; Edwards [3], Section 12.6; Hewitt and Ross [1], §19. In particular, if $f \in C(G)$ and $\nu \in M(G)$, then $\nu * f$ and $f * \nu$ are the elements of $C(G)$ defined by

$$\nu * f(x) = \int f(y^{-1}x)d\nu(y)$$

and

$$f * \nu(x) = \int f(xy^{-1})d\nu(y) ,$$

respectively, for every $x \in G$. If $\nu = \mu^g$ in the sense explained in 2.1.4(iii), then $\nu * f$ and $f * \nu$ agree with the functions $g * f$ and $f * g$, respectively, defined a little earlier via (2.1.2).

2.1.6. Central functions. By a <u>central function</u> is meant a function $k \in L^1(G)$ such that

$$k * f = f * k \qquad (2.1.8)$$

for every $f \in L^1(G)$ (or, equivalently, for every $f \in C(G)$). A continuous function k proves to be central if and only if

$$k(x) = k(y^{-1}xy) \text{ for every } x, y \in G , \qquad (2.1.9)$$

i.e., if and only if k is invariant under every inner automorphism of G.

For example, if $f \in C(G)$, then

$$k(x) = \int f(y^{-1}xy)dy$$

is a continuous central function.

As will appear in 2.2.6, the character of any continuous finite-dimensional representation of G is a continuous central function. Later (see 2.6.7, 2.9.2 and 2.9.3) it will emerge that every central function can be built up as a limit of finite sums of such characters.

Evidently, if G is Abelian, every integrable function is central.

2.1.7. Approximate identities. It has been noted in 2.1.5 that each of C(G) and $L^p(G)$ is a convolution algebra. Unless G is discrete, non of these algebras contains an identity, i.e., an element u such that $u * f = f * u = f$ for every f in the algebra. (On the contrary, M(G) does possess an identity for convolution, namely, the Dirac measure placed at the identity element of G.) However, so-called approximate identities exist in abundance and prove to be very useful.

By an approximate identity (in $L^1(G)$) is meant a sequence or net (k_j) of integrable functions on G such that

$$\left.\begin{aligned} &\sup_j \|k_j\|_1 < \infty, \\ &\lim_j \int k_j d\mu = 1, \\ &\lim_j \int_{G\setminus N} |k_j| d\mu = 0 \text{ for every neighbourhood } N \text{ of } e \text{ in } G. \end{aligned}\right\} \tag{2.1.10}$$

If G is metrisable (i.e., first countable), only sequences need be considered; otherwise nets will be needed.

The existence of approximate identities is settled by the following result, which will find use in 2.9.2 and 2.9.6.

2.1.8. Let (N_j) be a base of neighbourhoods of the identity element e in G, the index set being directed in such a way that $j \geq j_0$ implies $N_j \subseteq N_{j_0}$. There exists an approximate identity (k_j) in which each k_j is continuous, non-negative, centra, k_j vanishes on $G\setminus N_j$ and (see 2.8.1) k_j is positive definite.

Proof. First choose for each j an open symmetric neighbourhood of e, say N'_j, satisfying $N'_jN'_j \subseteq N_j$. Then (see Hewitt and Ross [1], (4.9)) choose an open symmetric neighbourhood N''_j such that $y^{-1}N''_jy \subseteq N'_j$ for every $y \in G$. (Here, as elsewhere, if A, B, ..., C denote subsets of a group G, AB ... C denotes the set of all group elements of the form ab ... c, where $a \in A$, $b \in B$, ... $c \in C$.) By a well-known result in general topology, there exists for each j a continuous function $h_j \geq \underline{0}$ on G, $h_j \neq \underline{0}$, h_j zero on $G\setminus N'_j$. Since N''_j is symmetric, one may suppose that $h_j = \check{h}_j$. Furthermore, in view of 2.1.2(iv), one may suppose that $\int h_j d\mu = 1$. Put

$$u_j : x \mapsto \int h_j(y^{-1}xy)dy ;$$

then $u_j \in C(G)$, $u_j \geq \underline{0}$, u_j is central (see 2.1.6), $\check{u}_j = u_j$, $\int u_j d\mu = 1$, and u_j vanishes on $G\backslash N_j'$. Finally, define $k_j = u_j * \check{u}_j = u_j * u_j$. It is simple to check that $k_j \in C(G)$, $k_j \geq \underline{0}$, k_j vanishes on $G\backslash N_j$ and $\int k_j d\mu = 1$; in particular, (k_j) is an approximate identity. That k_j is also positive definite appears (see 2.8.1) from the fact that it is of the form $u_j * \tilde{u}_j = u_j * \check{\bar{u}}_j$.

2.1.9. The utility of approximate identities, as well as the reason for the name, hinges on the following facts (see also Exercise 2.3.5).

Let (N_j) and (k_j) be as in 2.1.8. Then

(i) $\lim_j k_j * f(x) = \lim_j f * k_j(x) = f(x)$ for every $f \in L^1(G)$ and every point of continuity x of f; the convergence is uniform with respect to $x \in E$, whenever E is a closed subset of G containing only points of continuity of f.

Moreover, if E denotes any one of $C(G)$ and $L^p(G)$, where $1 \leq p < \infty$, then

(ii) if (k_j) is any approximate identity,

$$\lim_j k_j * f = \lim_j f * k_j = f$$

in the sense of the norm on E for every $f \in E$.

The proofs given in Edwards [3], 3.2.2 are easily adapted to cover the present case.

Note that (ii) is false when $E = L^\infty(G)$ or $M(G)$. Also, in case (i), $k_j * f = f * k_j \in C(G)$ for every j and every $f \in L^1(G)$.

2.1.10. Closed invariant subspaces. Denote by E any one of $C(G)$ or $L^p(G)$. As remarked in 2.1.4(iv), each L_a and each R_a may be viewed as a linear isometry of E onto itself. The type of question which leads into what has come to be lumped together as harmonic analysis is that which asks about the existence and structure of closed subspaces of E which are left (or right, of bi-) invariant, and about the possibility of synthesising E as some sort of direct sum of minimal closed invariant subspaces. What is placed under the heading of harmonic analysis is thus in part really a question of harmonic synthesis; see 2.13 below and Edwards [3], 2.2.1.

It turns out that, thanks to compactness of G, fairly definitive answers to this type of question are available. The problems are not trivial by any means (provision of some of the answers will occupy the remainder of these notes); on the other hand, analogous problems for locally compact non-compact groups are very much more difficult and the answers are to date much less complete.

These considerations, coupled with the easily-verified formulae

$$L_a L_b = L_{ab}, \quad R_b R_a = R_{ab} \quad (a, b \in G), \tag{2.1.11}$$

also serve to suggest the approach to be adopted in these notes. Thus, (2.1.2) indicates that each of $a \mapsto L_a$ and $a \mapsto R_{a^{-1}}$ are instances of (generally infinite dimensional) representations of G with representation space E (see 2.2 below). There is therefore a vague suggestion that the discussion of representations may be a useful tool. More specifically, one may hope to be able to break up these representations of G into more manageable components by restricting the L_a and R_a to subspaces M of E which are left- and right-invariant and as small as possible (without being collapsed to $\{0\}$ of course!). It does indeed turn out (see 2.12.14) that the only representations one has need to consider are in fact equivalent to components of this sort.

In this way, the traditional approach mentioned in 2.0.2(i) is suggested, and it is time to consider representations in a little more detail.

2.1.11. Exercise. (a) Describe the invariant integral of a general finite group.

(b) Verify in detail that the normalised invariant integral of the circle group **T** (see 2.0.4(ii)) is given by

$$f \in C(\mathbf{T}) \mapsto (2\pi)^{-1} \int_0^{2\pi} f(e^{it})dt ,$$

where the integral on the right is an ordinary Riemann integral.

2.1.12. Exercise. Suppose that G_1 and G_2 are compact groups with normalised invariant integrals I_1 and I_2. Show that the normalised invariant integral I of the product group $G = G_1 \times G_2$ is $I = I_1 \otimes I_2$, defined by

$$I(f) = \int \left(\int f(x_1, x_2) d\mu_{G_2}(x_2) \right) d\mu_{G_1}(x_1)$$

$$= \int \left(\int f(x_1, x_2) d\mu_{G_1}(x_1) \right) d\mu_{G_2}(x_2)$$

for every $f \in C(G)$.

Generalise to the product of any finite family of compact groups.

2.1.13. Exercise. Let G and G' be compact groups and h a continuous group homomorphism of G onto G'. Prove that

(i) $\int (f \circ h) d\mu_G = \int f d\mu_{G'}$

for every $f \in C(G')$.

Remark. In Bourbaki's language, (i) signifies that $\mu_{G'}$ is the h-image of μ_G. More generally, if X and Y are compact Hausdorff spaces, μ a Radon measure on X (see 1.10.6) and h a suitable (e.g., continuous) function from X into Y, one obtains a Radon measure ν on Y, termed the h-image of μ, by defining

(ii) $\int g d\nu = \int (g \circ h) d\mu$

for every $g \in C(Y)$. It is natural to expect that (ii) continues to hold for certain discontinuous complex-valued functions g on Y, and the study of this question is important; an exhaustive discussion appears in Bourbaki [2], Chapter 5.

2.1.14. Exercise. Let $(G_i)_{i \in I}$ be a family of compact groups, $G = \Pi_{i \in I} G_i$ the product group (see 2.0.4(iv)). For each finite subset F of I, let $G_F = \Pi_{i \in F} G_i$; and let $p_F : G \to G_F$ be defined by $(p_F x)_i = x_i$ for $i \in F$, where $x = (x_i)_{i \in I} \in G$. Similarly, if $F \subseteq F'$, let $p_{F'F} : G_{F'} \to G_F$ be defined by $(p_{F'F} y)_i = y_i$ for $i \in F$, where $y = (y_i)_{i \in F'} \in G_{F'}$. Let I_F denote the normalised invariant integral on G_F (see Exercise 2.1.12).

Write E for the linear subspace of C(G) formed of those $f \in C(G)$ such that $f = f_F \circ p_F$ for some finite f-dependent set $F \subseteq I$ and some $f_F \in C(G_F)$. Verify (by using the preceding exercise, for example) that $I_F(f_F)$ depends only upon f (and not upon F) and that $I: f \mapsto I_F(f_F)$ is a positive linear functional on E.

Verify that E is everywhere dense in C(G). (This may be done by judicious use of (2.1.7), 2.1.8 and 2.1.9; or by appeal to the Stone-Weierstrass theorem, for which see Edwards [2], p. 210.)

Deduce that I has a unique continuous extension to C(G) which is the normalised invariant integral of G.

2.1.15. **Exercise.** Let G be a compact group. For $f \in C(G)$, let $\check{f}: x \mapsto f(x^{-1})$ and $\tilde{f} = (\check{f})^- = (f^-)^\vee$, where the bar denotes complex conjugation. Verify that

$$f * g(e) = g * f(e), \quad (f * g)^\vee = \check{g} * \check{f},$$
$$(f * g)^- = \bar{f} * \bar{g}, \quad (f * g)^\sim = \tilde{g} * \tilde{f}$$

whenever $f, g \in C(G)$.

Formulate and prove generalisations of these formulae applying to more general functions f and g.

2.1.16. **Exercise.** Suppose that G is an infinite compact group with identity element e. Show that there exist open sets W_n $(n = 1, 2, \dots)$ such that $e \in \overline{W}_{n+1} \subseteq W_n$ and

$$0 < \mu_G(\overline{W}_{n+1}) < \tfrac{1}{2}\mu_G(W_n)$$

for every n. Deduce that $L^p(G)$ is infinite dimensional for every $p \in [1, \infty]$.

Let $F: [0, \infty) \to [0, \infty)$ be such that $t^{-1}F(t) \to \infty$ as $t \to \infty$, $F(0) = 0$. Show how to construct non-negative lower semicontinuous functions f on G such that f is integrable while $F \circ f$ is non-negative lower semicontinuous and non-integrable.

2.1.17. **Exercise.** Let Q denote the subgroup of the circle group **T** (see 2.0.4(ii)) composed of the 'rational points' $e^{2\pi it}$, where t ranges over all rational real numbers. Take Q with the topology induced by that of $\mathbf{T}$. Verify that Q is a (Hausdorff) Abelian topological group which is not locally compact (i. e., in which there exists no compact neighbourhood of the identity in Q).

Do there exist any nonzero invariant Borel measures μ on Q? (Invariance of μ means that $\mu(aA) = \mu(A)$ for every $a \in Q$ and every Borel subset A of Q; cf. 2.1.4(i).) Justify your answer.

Remark. Although Q is not locally compact, a fortiori not compact, it is in a sense not outrageously far removed from compactness: it is precompact (= totally bounded) in the sense that, for any neighbourhood U of the identity in Q, a finite number of translates of U suffice to cover Q (equivalently, the completion of Q ... which is none other than $\mathbf{T}$ ... is compact).

If one allows ∞ as a possible value of a measure, there are nonzero invariant Borel measures μ on Q, but these are pretty useless because $\mu(U) = \infty$ for every non-void open subset U of Q, and so $\underline{0}$ is the only continuous real or complex-valued function which is integrable.

2.1.18. **Exercise.** Suppose that G is a compact group and that $f \in C(G)$. Write $\Gamma(f)$ for the convex envelope in $C(G)$ of $\{L_a f : a \in G\}$ and $\overline{\Gamma}(f)$ for the closure in $C(G)$ of $\Gamma(f)$. Assuming the existence portions of statements 2.1.2(i)-(iii) and writing $c = I(f)$, show that $\underline{c} \in \overline{\Gamma}(f)$. More precisely, let ε be a given positive number; choose a neighbourhood W of e in G such that $x^{-1}x' \in W$ implies $|f(x) - f(x')| \leq \varepsilon$. Then choose (Hewitt and Ross [1], (4.9)) a neighbourhood V of e in G such that $x^{-1}VV^{-1}x \subseteq W$ for every $x \in G$. Express G as the union of measurable sets M_i $(i = 1, 2, \ldots, n)$ which are pairwise disjoint and such that $M_i \subseteq Vs_i$ for each i and some $s_i \in G$. Select $a_i \in M_i$ for each i and, by considering

$$\int_G f(y^{-1}x)\,d\mu(y) ,$$

show that the supremum norm of

$$\underline{c} - \sum_{i=1}^n \mu(M_i) L_{a_i} f$$

is at most ε.

Remark. Define the oscillation of any complex-valued function g on G to be $\sup\{|z - z'| : z, z' \in \text{Ran } g\}$. The above shows that $\Gamma(f)$ contains functions having arbitrarily small oscillation, and that $\overline{\Gamma}(f)$ contains a function whose oscillation is zero, such a function being $\underline{c}$, where $c = I(f)$. Pontryagin's proof of the existence of I, mentioned in 2.1.2, hinges on using Ascoli's theorem (Edwards [2], Corollary 0.4.12) to show that $\overline{\Gamma}(f)$ is compact; deducing from this the existence of $f_0 \in \overline{\Gamma}(f)$ which minimises the oscillation; and showing that f_0 is necessarily unique and a constant function. (It suffices to do this for $f \in C_{\mathbf{R}}(G)$.)

2.2. Group representations

These notes will not satisfy any interest in representations per se, being concerned solely with the way in which representations of compact groups enter into and act as tools in the study of harmonic analysis on such groups. In this section it is hoped to say almost all that need be said for the limited aims of what follows.

2.2.1. Basic definitions. If G is a group (not necessarily topologised for the moment), by a (finite-dimensional linear) representation of G is meant a homomorphism

$$U : x \mapsto U(x) \tag{2.2.1}$$

of G into the group $GL(\mathcal{H}_U)$ of invertible endomorphisms of some finite-dimensional nonzero linear space $\mathcal{H}_U$ onto itself; $\mathcal{H}_U$ is termed the representation space of (2.2.1), and the dimension of $\mathcal{H}_U$ is spoken of as the dimension or degree of (2.2.1).

Henceforth the phrases 'finite dimensional' or 'finite dimension' will be abbreviated to 'f.d.'.

A shade more precision should perhaps be offered at this point by remarking that algebraists often find it desirable to work with representations in which the representation space is a linear space over a general field F. In these notes, however, there will be no call to consider cases other than those in which F is either $\mathbf{R}$ (the real field) or $\mathbf{C}$ (the complex field), in which case one speaks of real or complex representations respectively. Moreover, once this restriction is made, there is no real loss in dealing only with complex representations. If a real representation U is given, having as representation space the real linear space V, one may always (see Halmos [2], p. 150) form a 'complexification' V^+ of V, which is a complex linear space into which V may be injected as a real-linear subspace so that V^+ is the direct sum of V and iV, and so that each U(x) is extendible into an invertible endomorphism $U^+(x) : \underset{\sim}{u} + i\underset{\sim}{v} \mapsto U(x)\underset{\sim}{u} + iU(x)\underset{\sim}{v}$ of V^+ which leaves V and V^+ invariant. What is more, if V is a real Hilbert space, its scalar product may be extended to V^+ so as to make the latter into a complex Hilbert space; one simply defines for $\underset{\sim}{u}_1, \underset{\sim}{u}_2, \underset{\sim}{v}_1, \underset{\sim}{v}_2 \in V$

$$(\underset{\sim}{u}_1 + i\underset{\sim}{v}_1 | \underset{\sim}{u}_1 + i\underset{\sim}{v}_2) = (\underset{\sim}{u}_1 | \underset{\sim}{u}_2) + (\underset{\sim}{v}_1 | \underset{\sim}{v}_2) - i(\underset{\sim}{u}_1 | \underset{\sim}{v}_2) + i(\underset{\sim}{v}_1 | \underset{\sim}{u}_2) ,$$

where, on the right, $(|)$ denotes the given inner (or scalar) product on V. If the original U(x) are unitary on V, the $U^+(x)$ are unitary on V^+.

In view of this, the term 'representation' will henceforth be taken to mean 'complex representation', unless the contrary is explicitly indicated.

It should also be pointed out that infinite dimensional representations are defined in a similar fashion, though in this case it is usual to assume in (2. 2. 1) that $\mathscr{H}_U$ is (at least) a topological linear space and that each U(x) is (at least) a linear homeomorphism of $\mathscr{H}_U$ onto itself. In the f. d. case, these extra topological conditions are automatically fulfilled when $\mathscr{H}_U$ is endowed with its unique Hausdorff linear space topology; see Edwards [5], Sections 1. 1 and 1. 2.

2. 2. 2. Continuity, measurability and boundedness of representations. When G is a topological group, the representation (2. 2. 1) is said to be <u>continuous</u> if and only if every <u>coordinate function</u>

$$x \mapsto f(U(x)\underset{\sim}{u}) , \qquad (2. 2. 2)$$

where $\underset{\sim}{u} \in \mathscr{H}_U$ and f is a continuous linear functional on $\mathscr{H}_U$, is continuous. If G is compact (or locally compact), so that a good invariant integral exists, the representation is termed <u>measurable</u> if and only if every one of these coordinate functions is measurable. There are similar definitions of continuity and integrability of representations. Whether or not G is topologised, the representation is said to be bounded if and only if every coordinate function is bounded. Cf. Hewitt and Ross [1], (22. 2) and (22. 8); Weil [1], Section 18.

These definitions apply whether or not the representation is f. d.

2. 2. 3. Unitary representations. In both the finite and infinite dimensional cases, the most tractable type of representation is that said to be unitary: by this it is meant that the representation space $\mathscr{H}_U$ is a Hilbert space and that every U(x) is a unitary endomorphism of $\mathscr{H}_U$. For example, the representation $a \mapsto L_a$ $(a \mapsto R_{a^{-1}})$ is unitary when considered as acting on any closed left (right) invariant subspace of $L^2(G)$.

For the case of compact groups G, it turns out that f. d. continuous unitary representations suffice for all that is required; see 2. 2. 8 below. (Continuous unitary representations, though usually infinite dimensional ones, are enough for locally compact groups... though these notes will neither prove nor utilise this fact.)

For a unitary U, the coordinate functions (2. 2. 2) are precisely the functions of the form $x \mapsto (U(x)\underset{\sim}{u} | \underset{\sim}{v})$, where $\underset{\sim}{u}, \underset{\sim}{v} \in \mathscr{H}_U$; cf. Halmos [2], §67. (The infinite dimensional analogue is true for continuous linear functions; see e. g. Edwards [2], 1. 12. 6.) However, it is slightly more convenient (and makes no ultimate difference - see 2. 7. 5) to take the coordinate functions in the form $x \mapsto (\underset{\sim}{u} | U(x)\underset{\sim}{v})$.

The reader must at this stage familiarise himself with some of the basic facts about f. d. Hilbert spaces and their endomorphisms. A good and very readable reference is Halmos [2], especially Chapter III. Note that the f. d. complex Hilbert spaces, which are especially relevant in what follows, are termed 'unitary spaces' in Halmos [2]. Additional points of detail are handled in Appendix A.

In what follows, inner (or scalar) products will be written $(|)$, and A* will always denote the adjoint of the Hilbert space endomorphism A.

2. 2. 4. Equivalence of representations. No one representation of G, say U, can be expected to tell us much about the structure of G...

save, that is, in the exceptional case where the representation is faithful (i. e., where the homomorphism U is actually an isomorphism). A worthwhile advance takes place when one knows of a whole family of representations which, taken together, provides a faithful picture.

It is evident that the contribution to the total picture made by any one representation U is no more and no less than that made by any other representation, say V, which is equivalent to U in the sense that there is a linear homeomorphism A of $\mathscr{H}_U$ onto $\mathscr{H}_V$ such that

$$AU(x) = V(x)A \quad \text{for every } x \in G . \qquad (2.2.3)$$

In the f. d. case, it is enough to demand merely that A is a linear space isomorphism of $\mathscr{H}_U$ onto $\mathscr{H}_V$ and that (2.2.3) holds.

If U and V are unitary representations of G, they are said to be unitarily equivalent if and only if (2.2.3) holds for some choice of A which is a linear isometry of $\mathscr{H}_U$ onto $\mathscr{H}_V$ (in which case A may itself be termed unitary, since it preserves scalar products as well as norms).

2.2.5. **Reducibility of representations.** The representation (2.2.1) is termed irreducible if and only if there exist no closed linear subspaces of $\mathscr{H}_U$ other than $\{0\}$ and $\mathscr{H}_U$ itself, which are invariant (i. e., stable) under every U(x); otherwise, the representation is said to be reducible. (This definition is framed so as to apply to infinite dimensional representations; in the f. d. case, the adjective 'closed' is superfluous and may be dropped.)

It is evident that two equivalent representations are together reducible or not.

For general (even general f. d.) representations U, this concept is not immediately very helpful. The reason is that, even if U is reducible, so that there is a nontrivial closed subspace M of $\mathscr{H}_U$ which is left invariant by every U(x), it does not follow that M has a supplementary closed subspace N (such that $\mathscr{H}_U$ is the direct sum $M \oplus N$) which also is invariant under every U(x). For unitary representations, however, the situation is very agreeable: the orthogonal complement $M^\perp$ does the trick: $\mathscr{H}_U$ is the direct sum of M and $M^\perp$, and $M^\perp$

is automatically invariant whenever M is (because every U(x) is unitary). Accordingly, $U': x \mapsto U(x)|M$ and $U'': x \mapsto U(x)|M^\perp$ are unitary representations and U decomposes in an obvious fashion into the 'sum' $U' \oplus U''$. Either or both of U' and U" may be reducible and the procedure repeatable. In the f. d. case, each of U' and U" has dimension less than that of U, and a simple inductive argument on the dimension of U now shows that $\mathcal{H}_U$ may be decomposed into a finite direct sum of pairwise orthogonal minimal invariant subspaces M_k, and U is correspondingly decomposed into a finite sum of f. d. irreducible unitary representations $U_k: x \mapsto U(x)|M_k$.

The preceding argument extends to continuous f. d. representations of compact groups, thanks to 2. 2. 8(b) below.

2. 2. 6. Traces and characters. Let $\mathcal{H}$ denote any f. d. complex linear space. The reader is referred to Appendix A. 2 for the definition and essential properties of the trace function with domain End($\mathcal{H}$). It suffices for the moment to depend on the following way of evaluating this trace function, denoted by Tr: take any base $(\underset{\sim}{a}_i)$ for $\mathcal{H}$ and let (f_i) denote the dual base for the space $\mathcal{H}$ of linear forms on $\mathcal{H}$ (determined by the conditions $f_i(\underset{\sim}{a}_j) = \delta_{ij}$); then

$$\mathrm{Tr}\ T = \sum_i f_i(T\underset{\sim}{a}_i) \qquad (2.2.4)$$

for every $T \in \mathrm{End}\,(\mathcal{H})$.

From (2. 2. 4) it is evident that Tr is a linear functional on End ($\mathcal{H}$). Furthermore (see Exercise 2. 2. 10)

$$\mathrm{Tr}\ T_1T_2 = \mathrm{Tr}\ T_2T_1 \qquad (2.2.5)$$

whenever $T_1, T_2 \in \mathrm{End}\,(\mathcal{H})$; and, if $\mathcal{H}'$ is also a f. d. complex linear space and A a linear space isomorphism of $\mathcal{H}$ onto $\mathcal{H}'$, then

$$\mathrm{Tr}\ ATA^{-1} = \mathrm{Tr}\ T\ . \qquad (2.2.6)$$

In case $\mathcal{H}$ is a Hilbert space, it is usual to specialise (2. 2. 4) into

$$\mathrm{Tr}\ T = \sum_i (T\underset{\sim}{e}_i | \underset{\sim}{e}_i)\ , \qquad (2.2.4')$$

where $(\underset{\sim}{e}_i)$ is any selected orthonormal base in $\mathscr{H}$. In this case,

$$\mathrm{Tr}\ T^* = \overline{\mathrm{Tr}\ T}. \qquad (2.2.7)$$

Reverting to the main theme, let U be a f.d. representation of the group G. The complex-valued function

$$\chi_U : x \mapsto \overline{\mathrm{Tr}\ U(x)} \qquad (2.2.8)$$

is termed the <u>character</u> of U. It is more usual to term $\overline{\chi}_U$ the character of U, but the definition (2.2.8) is slightly more convenient for the purposes of these notes.

From (2.2.4) it appears that χ_U is a finite sum of coordinate functions; hence χ_U is bounded (or measurable, or continuous, or integrable) whenever U has the same property. As a consequence of (2.2.6),

$$\chi_U(axa^{-1}) = \chi_U(x) \qquad (2.2.9)$$

for every $a, x \in G$; hence, if U is integrable, χ_U is a central function (see 2.1.6).

If U is unitary, (2.2.7) shows that

$$\chi_U(x^{-1}) = \overline{\chi_U(x)} \qquad (2.2.10)$$

for every $x \in G$, i.e., that $\chi_U = \overline{\chi}_U$ and $\tilde{\chi}_U = \chi_U$ (see Exercise 2.1.15). Furthermore, (2.2.4') leads to

$$|\chi_U(x)| \le \chi_U(e) = d(U) \qquad (2.2.11)$$

for every $x \in G$.

It is a consequence of (2.2.6) that equivalent representations have the same character. Much less obvious (but a corollary of 2.2.8(b) and the results mentioned in 2.6.5) is the fact that, if two continuous representations of a compact group G have the same character, then they are equivalent. This circumstance is partly responsible for the choice of the term 'character'.

2.2.7. One dimensional representations. The case of a one-dimensional representation is especially noteworthy. Here one has $U(x) = \overline{\chi(x)} I_U$, where I_U denotes the identity endomorphism of $\mathscr{H}_U$ and where χ is a complex-valued function on G such that

$$\chi(x) \neq 0, \quad \chi(xy) = \chi(x)\chi(y) \tag{2.2.12}$$

for every $x, y \in G$. Evidently, χ_U is now none other than χ. If the representation is bounded (in particular if it is unitary), (2.2.12) implies that

$$|\chi(x)| = 1 \quad (x \in G) ; \tag{2.2.13}$$

in other words, χ is a homomorphism of G into **T**. Conversely, if (2.2.12) holds, then $x \mapsto \chi(x)I_U$ yields a representation on any space $\mathscr{H}_U$ on which I_U denotes the identity endomorphism; this representation is irreducible if and only if $\dim \mathscr{H}_U = 1$; and, if (2.2.13) holds, this representation is unitary on any Hilbert space $\mathscr{H}_U$ on which I_U denotes the identity endomorphism.

A complex-valued function χ on a group G which satisfies (2.2.12) is termed a <u>multiplicative character</u> of G; as has been seen, multiplicative characters of G may be identified with the characters of one-dimensional representations of G. A bounded multiplicative character of G satisfies (2.2.13) and may be thought of as the character of some one-dimensional unitary representation of G.

Multiplicative characters prove to be especially important in the case of Abelian groups, the principal reason being that (as will appear in 2.5.4) every f.d. irreducible representation of an Abelian group is one-dimensional and may be identified with its (multiplicative) character. (This is valid too for continuous irreducible unitary representations of arbitrary dimension of locally compact Abelian groups, but this will not be established in these notes.) When this fact is combined with the completeness theorem (CTii) in 2.4.1, it will appear that every compact Abelian group G has 'sufficiently many' continuous multiplicative characters (of unitary representations), i.e, if x and y denote distinct elements of G, there is a continuous multiplicative character χ of G satisfying (2.2.13) and such that $\chi(x) \neq \chi(y)$.

On the other hand, there are compact non-Abelian groups having no non-trivial multiplicative characters at all; see Exercise 2. 2. 12.

In view of 2. 2. 8(a), discontinuous characters are extremely ill-behaved. They do exist, in general; see Exercise 2. 2. 14.

2. 2. 8. **Recital of basic facts.** The following summary of some basic facts is included as an aid to the reader in evaluating the situation, and by way of explanation of the adequacy for our purposes of looking only at continuous f. d. unitary representations of compact groups. No explicit use of (a) or (d) will be made in the sequel; (e) will be used but twice (in 2. 7. 5); (b) will be, and (c) has been, dealt with; and (f), which is absolutely vital to us, will be discussed in detail in 2. 4 and 2. 8. 8 below.

(a) Every f. d. measurable representation is continuous (Hewitt and Ross [1], (22. 19; Weil [1], §18).

(b) Every continuous f. d. representation is equivalent to a continuous f. d. unitary representation (Exercise 2. 2. 11 below; Weil [1]), §19; Pontryagin [1], p. 110; Naimark [1], p. 431).

(c) Every f. d. unitary representation can be decomposed into a finite sum of irreducible f. d. unitary representations, each summand being continuous if the given representation is continuous (see 2. 2. 5 above).

(d) Every continuous irreducible unitary representation is f. d. (Hewitt and Ross [1], (22. 13); Naimark [1], p. 434).

(e) Any two equivalent f. d. irreducible unitary representations are unitarily equivalent (Hewitt and Ross [1], (27. 13)).

(f) The completeness theorem: the coordinate functions $x \mapsto (\underset{\sim}{u} | U(x)\underset{\sim}{v})$ corresponding to all continuous f. d. irreducible unitary representations of G form a complete subset of $L^2(G)$ (see 2. 4 below).

In (b), (d) and (f) it is essential that G be compact (see, for example, Hewitt and Ross [1], (22. 20. c) and (22. 22)). There are infinite dimensional versions of (c), but they are relatively tricky and complicated; see, for example Hewitt and Ross [1], (27. 44) and the remarks on p. 354 of Volume I referring to (22. 13). (e) is valid for any group.

We might add here that a detailed discussion of the invariant integral and representations of specific non-Abelian compact groups (such as SU(n) and SO(n) is a complex and decidedly non-trivial task; see 2.2.17 below.

2.2.9. Notation; the set $\hat{G}$. The following notation will be used from this point onward, unless the contrary is explicitly indicated for each temporary digression.

G will denote a Hausdorff compact group with elements x, y, ... and identity element e. In general, G will be multiplicatively written. The normalised bi-invariant integral on G will (see 2.1.4) be indicated in bound variable notation by

$$\int_G \ldots dx \quad \text{or} \quad \int \ldots dx ;$$

$L^p(G)$ will denote the usual Lebesgue space formed with respect to the chosen bi-invariant measure on G, the norm on $L^p(G)$ being denoted by $\|.\|_p$. C(G) will denote the Banach space of continuous complex-valued functions on G, and M(G) the space of complex Radon measures on G (see 1.7.9 and 1.11).

It will be necessary to introduce an object which is in some sense 'dual' to G: a possible contender is the set $\hat{G}$ constructed in the following manner. Given G, form the set of all f.d. continuous irreducible unitary representations of G. (In order to skirt any set-theoretical difficulties here, it may be noted that each of the f.d. representation spaces involved could be replaced by a subspace of the fixed Hilbert space l^2 of square-summable complex sequences; when this is done, each corresponding representation is replaced by an equivalent one.) Partition this set modulo the relation of equivalence between representations, and suppose selected precisely one member of each equivalence class to obtain a set $\hat{G}$ of representations. This choice will be supposed made once for all.

The arbitrariness of this choice is one of the mildly dissatisfying features of this approach. A rather less abstract picture of $\hat{G}$ may be formed by indexing it by the set of continuous characters of G (which, unlike $\hat{G}$, is intrinsically related to G); but the possibility of doing this is not clear at the outset (see 2.6.5 below).

If $U \in \hat{G}$, the corresponding representation space will be denoted by $\mathscr{H}_U$, its dimension (a positive integer) by $d(U)$, and the identity endomorphism of $\mathscr{H}_U$ by I_U. The same conventions will apply to any f. d. continuous unitary representation U of G, whether or not $U \in \hat{G}$.

2.2.10. Exercise. Prove equations (2.2.5)-(2.2.7). (Regarding (2.2.6), note that if $(\underset{\sim}{a}_i)$ and (f_i) are dual bases for $\mathscr{H}$ and $\mathscr{H}^*$, then $(\underset{\sim}{a}_i') = (A\underset{\sim}{a}_i)$ and $(f_i') = (f_i \circ A^{-1})$ are dual bases for $\mathscr{H}'$ and $\mathscr{H}'^*$.)

2.2.11. Exercise. Let G be a compact group and $S: x \mapsto S(x)$ a continuous f. d. representation of G. Show that S is equivalent to a continuous unitary representation U of G.
[Hint: Let $\mathscr{H}$ denote the representation space of S. Show that one can define a scalar product $(\,|\,)_0$ making $\mathscr{H}$ into a Hilbert space $\mathscr{H}_0$. Then consider the new scalar product defined by

$$(\underset{\sim}{u}|\underset{\sim}{v}) = \int (S(x)\underset{\sim}{u}|S(x)\underset{\sim}{v})_0 dx\ .\ ]$$

2.2.12. (i) **Exercise.** It is known (see Macdonald [1], pp. 114, 220) that there exist non-trivial finite groups G which are simple, i. e., which have no normal subgroups other than $\{e\}$ and G. Show that any such group has no non-trivial multiplicative characters.

(ii) Let G_i $(i = 1, 2, \dots)$ be a non-trivial finite group having no non-trivial multiplicative characters. Regard each G_i as a compact group with its discrete topology, and let G be the produce of the G_i; G is an infinite compact group. Show that G has no non-trivial continuous multiplicative characters. (Note that G has many closed normal subgroups different from $\{e\}$ and G.)

2.2.13. Remark. Groups $G = \Pi_{i=1}^{\infty} G_i$, in which each G_i is finite and discrete, have a number of rather unfamiliar properties stemming from the fact that there is a base of neighbourhoods of e in G composed of closed subgroups H_j $(j = 1, 2, \dots)$ of G. (Compare Hewitt and Ross [1], (4.21) and (10.2).) When this is the case, each H_j is also open in G. As a result, G is totally disconnected (i. e., every non-void connected subset of G is a singleton) and zero-dimensional (loc. cit. (3.5)). If χ is a continuous multiplicative character of G, $\chi(H_j) = \{1\}$ for some j and so $\chi(G)$ is a finite subgroup of **T** and therefore $\chi^n = \underline{1}$ for some n (cf. Hewitt and Ross [1], (24.26)). Similarly, if U is any f.d. continuous representation of G, U(G) is a finite subgroup of $GL(\mathscr{H}_U)$. It turns out that groups G of this sort behave, in respect to convergence properties of Fourier series, quite differently from the familiar groups (such as **T** and its finite products); see the end of 2.9.7 below.

2.2.14. Exercise. Consider **R** as a linear space over the rational field **Q**. Show that one can choose $a_n \in \mathbf{R}$ $(n \in \mathbf{N} = \{1, 2, \ldots\})$ such that $a_1 = 2\pi$, $\lim_{n\to\infty} a_n = 0$ and the family $(a_n)_{n\in\mathbf{N}}$ is linearly independent over **Q**. Extend this family into a base $(a_m)_{m\in M}$ for **R** over **Q**, M being a suitable superset of **N**. Then every $x \in \mathbf{R}$ has a unique expression $x = \sum_{m\in M} \nu_m(x) a_m$, where $\nu_m(x) \in \mathbf{Q}$ and $\{m \in M : \nu_m(x) \neq 0\}$ is finite. Let $(b_m)_{m\in M}$ be an arbitrary family of complex numbers and define $\chi : \mathbf{R} \to \mathbf{C}$ by $\chi(x) = \prod_{m\in M} \exp(b_m \nu_m(x))$. Show that χ is a multiplicative character of **R** such that $\chi(a_m) = \exp b_m$ for every $m \in M$. If $b_1 = 0$, χ has period 2π and so defines a multiplicative character χ_1 of **T** via the formula $\chi_1(e^{it}) = \chi(t)$ for every $t \in \mathbf{R}$. Thus $\chi_1(e^{ia_n}) = \exp b_n$ for every $n \in \mathbf{N}$. Show how to arrange for χ_1 to be discontinuous.

2.2.15. Exercise. Show that the continuous multiplicative characters of the circle group **T** are the functions

$$\chi_n : e^{it} \mapsto e^{int},$$

where $n \in \mathbf{Z}$.

2.2.16. Exercise. Let $(\mathscr{H}_i)_{i\in I}$ be a family of Hilbert spaces and denote by $\mathscr{H}$ the subset of $\prod_{i\in I} \mathscr{H}_i$ formed of families $(\underset{\sim}{a}_i)$ such that

$$\sum_{i\in I} \|\underset{\sim}{a}_i\|^2 < \infty.$$

Verify that $\mathscr{H}$ is a Hilbert space relative to the scalar product

$$((a_i) \mid (b_i)) = \sum_{i \in I} (a_i \mid b_i) .$$

Show also that, if the $\mathcal{H}_i$ are pairwise orthogonal closed linear subspaces of a fixed Hilbert space $\mathcal{K}$, then $\mathcal{H}$ (as defined above) is isomorphic to the closed linear subspace of $\mathcal{K}$ generated by $\cup_{i \in I} \mathcal{H}_i$, this last subspace being taken with the structure induced by that of $\mathcal{K}$.

Note. $\mathcal{H}$ is usually termed the Hilbertian direct sum of the family $(\mathcal{H}_i)_{i \in I}$ and denoted by $\oplus_{i \in I} \mathcal{H}_i$. In case the $\mathcal{H}_i$ are pairwise orthogonal closed linear subspaces of a Hilbert space $\mathcal{K}$, we speak of an internal Hilbertian direct sum.

2.2.17. An example. Nowhere in these notes do we attend to the problem of finding sufficiently many continuous irreducible unitary representations of any specific non-Abelian compact group. As far as the writer is aware, there is indeed no general method for doing this, or for tackling many of the sub-problems involved. The interested reader should consult Hewitt and Ross [1], Volume II, pp. 152-6 for general comments and references. In order to give some idea of the techniques involved, we shall here describe briefly the procedure in one specific case; for all the details see loc. cit. (29.13)-(29.27).

The special case to be considered is the group $G = SU(2) = SU(V)$, where V is a complex Hilbert space of dimension 2; see 2.0.4(iii) above. Choosing a fixed orthonormal base in V, we regard elements x of G as 2×2 complex matrices,

$$x = \begin{bmatrix} \alpha & \beta \\ -\bar{\beta} & \bar{\alpha} \end{bmatrix}$$

satisfying $\det x = |\alpha|^2 + |\beta|^2 = 1$. For typographical reasons, we denote this element x by $\alpha \nabla \beta$. The aim is to obtain continuous irreducible unitary representations of G, perhaps enough of them to be sure that every such representation is equivalent to one of those obtained.

To this end, let M denote the set of non-negative integers and half-integers: $M = \{0, 1/2, 1, 3/2, 2, \ldots\}$. For each $m \in M$, denote by $\mathcal{H}_m$ the complex linear space of all polynomials f over the complex field in one indeterminate z and having degree at most $2m$. Make $\mathscr{H}_m$ into a complex Hilbert space by requiring that the monomials

$$((m-j)!\,(m+j)!)^{-\frac{1}{2}} z^{m-j} \qquad (j \in \{-m, -m+1, \ldots, m-1, m\})$$

form an orthonormal base in $\mathscr{H}_m$.

For each $m \in M$ and $x = \alpha \nabla \beta \in G$ define the endomorphism $U^{(m)}(x)$ of $\mathscr{H}_m$ to be such that

$$U^{(m)}(x) z^s = (\beta z + \overline{\alpha})^{2m-s} (\alpha z - \overline{\beta})^s$$

for every $s \in \{0, 1, 2, \ldots, 2m\}$. A rather elaborate calculation will verify that then

$$U^{(m)} : x \mapsto U^{(m)}(x)$$

is a continuous unitary representation of G.

To show that each $U^{(m)}$ is irreducible, we introduce the following continuous homomorphisms w_1 and w_2 of $\mathbf{R}$ into G:

$$w_1(t) = \cos(\tfrac{1}{2}t) \nabla i \sin(\tfrac{1}{2}t), \qquad w_2(t) = \cos(\tfrac{1}{2}t) \nabla -\sin(\tfrac{1}{2}t)$$

and then the associated so-called infinitesimal operators of the representation $U^{(m)}$:

$$A_k^{(m)} = (U^{(m)} \circ w_k)'(0),$$

$$H_+^{(m)} = iA_1^{(m)} - A_2^{(m)}, \qquad H_-^{(m)} = iA_1^{(m)} + A_2^{(m)}.$$

It is possible to verify that

$$H_+^{(m)} f = -f', \qquad H_-^{(m)} f = -2mzf + z^2 f'. \tag{2.2.14}$$

Now, if S is a linear subspace of $\mathscr{H}_m$ which is invariant under $U^{(m)}$, then S is also invariant under the $A_k^{(m)}$, and hence under the $H_+^{(m)}$ and $H_-^{(m)}$ and their iterates. Then, if $S \neq \{0\}$, (2.2.14) shows that S contains every monomial z^s with $s \in \{0, 1, 2, \ldots, 2m\}$, and hence that $S = \mathscr{H}_m$. This shows that $U^{(m)}$ is irreducible.

A further argument, too long to detail here, will show that every continuous irreducible unitary representation of G is equivalent to $U^{(m)}$ for some $m \in M$. Thus we might in this case take $\hat{G}$ to be $\{U^{(m)} : m \in M\}$.

Regarding the corresponding characters, note first that the spectral theorem (Appendix A. 1. 2) implies that to every $x \in G$ correspond $a \in G$ and $t \in \mathbf{R}$ such that

$$a^{-1}xa = w_3(t) = \exp(\tfrac{1}{2}it) \triangledown 0 .$$

So, by (2. 2. 9), every continuous character of G is fully determined by its restriction to the subgroup $w_3(\mathbf{R})$ of G. It turns out that χ_m, the character of $U^{(m)}$, is determined by the formula

$$\chi_m \circ w_3(t) = \begin{cases} \sin(m + \frac{1}{2})t/\sin(\frac{1}{2}t) & \text{if } \exp(\frac{1}{2}it) \neq \pm 1 \\ (2m + 1)(\exp(\frac{1}{2}it))^{2m} & \text{otherwise} \end{cases}$$

2. 3. The Fourier transform

2. 3. 1. For $f \in L^1(G)$ and $U \in \hat{G}$, write

$$\hat{f}(U) = \int_G f(x)U(x)dx , \qquad (2. 3. 1)$$

so that $\hat{f}(U) \in \text{End}\,(\mathscr{H}_U)$. The function $\hat{f}$ is termed the Fourier transform of f.

It is evident that $f \mapsto \hat{f}$ is linear and that

$$\|\hat{f}(U)\| \leq \|f\|_1 , \qquad (2. 3. 2)$$

the norm on the left being the standard operator norm on $\text{End}\,(\mathscr{H}_U)$; see (2. 0. 1).

In addition, if $\tilde{f} : x \mapsto (f(x^{-1}))^-$, then

$$(\tilde{f})\hat{}(U) = \hat{f}(U)^* \qquad (2. 3. 3)$$

for every $U \in \hat{G}$.

Introducing the convolution

$$f * g(x) = \int f(y)g(y^{-1}x)dy$$

as in 2. 1. 6, a direct calculation gives

$$(f * g)\hat{}(U) = \hat{f}(U)\hat{g}(U) \tag{2. 3. 4}$$

for every $U \in \hat{G}$. (Some use of the Fubini-Tonelli theorem is usually involved here.)

Furthermore,

$$(L_a f)\hat{}(U) = U(a)\hat{f}(U), \quad (R_a f)\hat{}(U) = \hat{f}(U)U(a) \tag{2. 3. 5}$$

for every $a \in G$ and every $U \in \hat{G}$.

In terms of characters one has

$$\begin{aligned} f * \chi_U(x) = \chi_U * f(x) &= \mathrm{Tr}[\hat{f}(U)U(x)*] \\ &= \mathrm{Tr}[U(x)*\hat{f}(U)] . \end{aligned} \tag{2. 3. 6}$$

Definition (2. 3. 1) extends in an obvious way to the case in which f is replaced by a measure $\nu \in M(G)$, and there are corresponding extensions of (2. 3. 2)-(2. 3. 6) which need not be written out explicitly.

This same definition obviously makes sense whenever U is any continuous f. d. representation of G (and even for suitably restricted infinite dimensional representations too). This possibility is used in the course of 2. 4. 2 below, and there only.

It is important to indicate that (2. 3. 1) differs slightly from the corresponding definition in Hewitt and Ross [1], (28. 34). However, the difference is purely formal in nature and is of no real significance.

2. 3. 2. The Abelian case. If G is Abelian then, as will be seen in detail in 2. 5. 4 below, every $U \in \hat{G}$ is one-dimensional and may be identified with its character, which is a continuous multiplicative character χ of G in the sense described in 2. 2. 7, i. e., a continuous homomorphism of G into $\mathbf{T}$. So, if we introduce the set Γ of all such continuous multiplicative characters of G, f appears the complex-valued function $\hat{f}$ on Γ:

$$x \mapsto f(\chi) = \int_G f(x)\overline{\chi(x)}dx . \tag{2. 3. 7}$$

Henceforth, whenever G is Abelian we shall assume that this change in stance is made.

By way of an 'aside' we might point out here that the compactness of G simplifies matters by arranging that the definition (2.3.6) or (2.3.7) makes good sense for every function f for which it is natural at the outset to expect a Fourier transform to be defined. If one passes to a noncompact Abelian group G, this is no longer true: the definition works perfectly well for functions $f \in L^1(G)$, but one soon finds the need for a definition of $\hat{f}$ applicable to certain non-integrable functions f. The customary first steps in easing the situation is to frame a good definition for functions $f \in L^2(G)$, and then to utilise a version of the Hausdorff-Young theorem discussed in 2.14 below to cope with functions f belonging to $L^p(G)$ for some $p \in (1, 2)$. (None of this is at all trivial: to go further still, something akin to distributional methods are needed.)

Starting, as we have done, from a compact G, this sort of problem arises only when one seeks to reconstitute f from $\hat{f}$, i.e., when one comes to study the convergence and summability of Fourier series of functions on G, as in 2.7 and 2.9. When it does so arise, the underlying group is Γ rather than G (Γ as a group is discussed in 2.5.4 below). As will be seen, these problems are quite delicate, and this despite the fact that compactness of G ensures yet another simplifying feature, namely, that Γ is discrete. The problems mentioned in the last paragraph are at their worst when the group (G or Γ) involved is neither compact nor discrete.

2.3.3. Remark. The formulae (2.3.2)-(2.3.4) assert that, for a fixed $U \in \hat{G}$, the mapping $f \mapsto \hat{f}(U)$ is a certain sort of representation of the convolution algebra $L^1(G)$ into the algebra $\mathrm{End}(\mathscr{H}_U)$. In some approaches to harmonic analysis, use is made of the fact that this passage from representations of G to representations of $L^1(G)$ is in some measure reversible. The details are fairly lengthy; see Hewitt and Ross [1], §§21 and 22, especially (22.7). The advantage gained by this device is due mainly to the fact that the algebra $L^1(G)$ is in some ways structurally 'richer' than the underlying group, and this extra richness can sometimes be exploited with effect (as it is in the Gelfand-Raikov approach mentioned in 2.0.2(iv)). The connection between Fourier transforms and representations of $L^1(G)$ is one reason for the basic importance of the former; the Abelian case is especially simple and striking (see Edwards [3], Chapter 4 and Edwards [4], §6).

2.3.4. Exercise. Write out detailed proofs of (2.3.2)-(2.3.5).

2.3.5. Exercise. Let (k_j) be an approximate identity in $L^1(G)$. Show that, for every $U \in \hat{G}$,

$$\lim_j \hat{k}_j(U) = I_U$$

in the sense of the operator norm on $\mathrm{End}(\mathscr{H}_U)$ (defined as in equation (2.0.1), with $\mathscr{H}_U$ in place of V).

2.3.6. Exercise. Discuss Γ and the Fourier transform in case G is a finite cyclic group.

2.4. The completeness and uniqueness theorems

2.4.1. What has come to be known as the <u>completeness theorem</u> may be understood to mean either of two assertions, namely:

(CTi) the set of coordinate functions

$$x \mapsto (\underset{\sim}{u} \,|\, U(x)\underset{\sim}{v}),$$

where $U \in \hat{G}$ and $\underset{\sim}{u}, \underset{\sim}{v} \in \mathscr{H}_U$, is complete in (i.e., generates a dense linear subspace of) $L^2(G)$;

(CTii) there exist sufficiently many continuous irreducible unitary representations of G, i.e., if $x, y \in G$ and $x \neq y$, then $U(x) \neq U(y)$ for at least one $U \in \hat{G}$.

Quite different in appearance from the completeness theorem (or theorems) are the following <u>uniqueness theorems,</u> one corresponding to each $p \in [1, \infty]$:

$$(\mathrm{UT}_p)\ f \in L^p(G),\ \hat{f}(U) = 0\ (\forall\, U \in \hat{G}) \Rightarrow f = \underline{0} \text{ a.e.}$$

In spite of appearances, however, the two types of theorem are essentially equivalent. The Hahn-Banach theorem coupled with known facts about the representation of continuous linear functionals on $L^2(G)$ (see, for example, Edwards [3], Appendices B.5 and C), shows that (CTi) and (UT_2) are equivalent. On the other hand, it is evident that (UT_1) implies

(UT_p) for every $p \in [1, \infty]$; and the use of approximate identities will show that the converse is also true (see Exercise 2.4.11). Since in fact (UT_1) is the assertion for which the need is most urgent, its proof will be the aim of this section (and will form perhaps the most substantial piece of analysis to be tackled in detail in these notes). The truth of (CTii) will be inferred later by showing it to be equivalent to (CTi); see 2.8.8 and 2.9.6.

2.4.2. Proof of (UT_1). The substance of 2.1.8 and 2.1.9 combines with (2.3.4) to show that all the statements (UT_p), obtained when p varies over $[1, \infty]$, are equivalent to the statement which results when $C(G)$ replaces $L^p(G)$ in the hypothesis. Moreover, since (2.3.3) and (2.3.4) combine to show that $(f * \tilde{f})\hat{} = \hat{f}\hat{f}^*$, it will suffice to show that

$$f \in C(G), \quad f = \tilde{f}, \quad \hat{f}(U) = 0 \quad (\forall\, U \in \hat{G}) \Rightarrow f = \underline{0}. \qquad (2.4.1)$$

The ensuing argument in fact shows that, if $f \in C(G)$ satisfies $f = \tilde{f} \neq \underline{0}$, then $\hat{f}(U) \neq 0$ for some $U \in \hat{G}$. In doing this, use will be made of elementary Hilbert space theory applied to $L^2(G)$ and the endomorphism T of $L^2(G)$ defined by

$$T : g \mapsto g * f. \qquad (2.4.2)$$

The following two properties of T are relevant and fairly simple to establish (see Exercise 2.4.6):

(i) $T \neq 0$ and T is self-adjoint, i.e. (see Appendix A.0),

$$(Tg \mid h) = (g \mid Th)$$

for every $g, h \in L^2(G)$;

(ii) T commutes with every left translation L_a $(a \in G)$.

A vital but less evident property is

(iii) T is compact (= completely continuous), i.e., if (g_j) is any bounded sequence in $L^2(G)$, there is a subsequence of (Tg_j) which converges in $L^2(G)$;

this will be established in 2.4.3. In 2.4.4 it will be shown that (i) and (iii) entail

(iv) T admits at least one non-zero eigenvalue λ, and the corresponding eigenmanifold

$$M = \{g \in L^2(G) : Tg = \lambda g\}$$

has finite positive dimension.

Actually, (iv) is a special case of a well-known general result stated in 2. 4. 4 as a lemma.

The proof of the uniqueness theorem will now proceed on the assumption that (i)-(iv) have been established.

Since λ is non-zero and $M \neq \{\underline{0}\}$, one may choose $g_0 \neq \underline{0}$ in M and infer that $g_0 = \lambda^{-1}(g_0 * f) \in C(G)$ and

$$g_0 * f = \lambda g_0 \neq \underline{0} .$$

Now (see Exercise 2. 4. 7) $f * L_a g_0(e) = g_0 * f(a^{-1})$ for every $a \in G$, and so a may be chosen so that

$$f * L_a g_0 \neq \underline{0} .$$

By (ii), M is invariant under all the L_a. Thus $g_1 = L_a g_0 \in M$ and therefore

$$f * g_1 \neq \underline{0} \quad \text{for some } g_1 \in M . \qquad (2.\ 4.\ 3)$$

Look now at the f. d. representation $V: x \mapsto V(x)$ of G with representation space M and defined by

$$V(x)g = L_x g .$$

It is simple to verify that V is continuous and unitary. Moreover, its Fourier transform

$$\hat{f}(V) = \int_G f(x)V(x)dx$$

turns out to be given by

$$\hat{f}(V) : g \mapsto f * g . \qquad (2.\ 4.\ 4)$$

Thus (2. 4. 3) shows that

$$\hat{f}(V) \neq 0 . \tag{2.4.5}$$

On the other hand, by 2.8.8(b) and the process indicated in 2.2.5, V may be decomposed into a finite sum of f.d. continuous irreducible representations of G. It is therefore equivalent to a sum $U_1 \oplus \ldots \oplus U_k$, where $U_1, \ldots, U_k \in \hat{G}$. Accordingly, $\hat{f}(V)$ is equivalent to

$$\hat{f}(U_1) + \ldots + \hat{f}(U_k) ,$$

and (2.4.5) plainly entails that $\hat{f}(U_j) \neq 0$ for some $j \in \{1, 2, \ldots, k\}$, as had to be established; cf. (2.4.1).

2.4.3. Proof of 2.4.2(iii). Sticking to the notation used in 2.4.2(iii), it suffices (by Ascoli's theorem; see for example, Edwards [2], 0.4.12) to show that the sequence (Tg_j) is equicontinuous and such that $\sup_j \|Tg_j\|_\infty < \infty$. (The compactness of G ensures that the injection map of C(G) into $L^2(G)$ is continuous.) However, by (2.1.5),

$$\|Tg_j\|_\infty \leq \|f\|_2 \|g_j\|_2$$

and $m = \sup_j \|g_j\|_2 < \infty$ by hypothesis. Also,

$$Tg_j(a) - Tg_j(b) = g_j * (R_{a^{-1}}f - R_{b^{-1}}f)(e) ;$$

by (2.1.5) again this is in absolute value at most

$$m \|R_{a^{-1}}f - R_{b^{-1}}f\|_2 = m \|R_{a^{-1}b}f - f\|_2 ,$$

the last step by invariance of the integral. Since f is continuous and G is compact, f is uniformly continuous. Hence, given $\varepsilon > 0$, there is a neighbourhood N of e in G such that $\|R_s f - f\|_2 \leq m^{-1}\varepsilon$ whenever $s \in N$. So $a^{-1}b \in N$ implies

$$|Tg_j(a) - Tg_j(b)| \leq \varepsilon$$

for every j, whence equicontinuity of (Tg_j).

2.4.4. Proof of 2.4.2(iv). This will be derived by specialisation from the following lemma about Hilbert spaces (see, for example, Edwards

[2], 9.11.2).

Lemma. Let $\mathcal{H}$ be a Hilbert space (of arbitrary dimension) and T a compact self-adjoint endomorphism of $\mathcal{H}$, $T \neq 0$. Then

(a) T admits at least one non-zero eigenvalue λ;

(b) for any non-zero eigenvalue λ of T, the associated eigen-manifold

$$M = \{\underset{\sim}{u} \in \mathcal{H} : T\underset{\sim}{u} = \lambda \underset{\sim}{u}\}$$

has positive finite dimension.

Proof. Since T is compact, the sequence $(T\underset{\sim}{u}_j)$ is certainly bounded whenever $(\underset{\sim}{u}_j)$ is a bounded sequence of elements of $\mathcal{H}$. From this it follows that T is continuous, i.e., that

$$m = \sup\{\|T\underset{\sim}{u}\| : \underset{\sim}{u} \in \mathcal{H}, \ \|\underset{\sim}{u}\| \leq 1\} < \infty . \tag{2.4.6}$$

Since T is self-adjoint, $(T\underset{\sim}{u}|\underset{\sim}{u})$ is real for every $\underset{\sim}{u} \in \mathcal{H}$. From (2.4.6) it follows that each of

$$\lambda = \sup\{(T\underset{\sim}{u}|\underset{\sim}{u}) : \|\underset{\sim}{u}\| \leq 1\}$$

and

$$\mu = \inf\{(T\underset{\sim}{u}|\underset{\sim}{u}) : \|\underset{\sim}{u}\| \leq 1\}$$

is finite. Since $T \neq 0$, at least one of λ or μ is non-zero; see Appendix A.0.6, A.0.8. By changing T into -T if need be, it may and will be assumed that

$$\lambda = \sup\{(T\underset{\sim}{u}|\underset{\sim}{u}) : \|\underset{\sim}{u}\| \leq 1\} > 0 . \tag{2.4.7}$$

The aim now is to show that λ is an eigenvalue of T, and the first step toward this is to show that λ is an assumed maximum, i.e., that

$$\lambda = (T\underset{\sim}{u}_0|\underset{\sim}{u}_0) \tag{2.4.8}$$

for some $\underset{\sim}{u}_0 \in \mathcal{H}$ satisfying

$$\|\underset{\sim}{u}_0\| = 1 . \tag{2.4.9}$$

To this end, note first that (2. 4. 7) guarantees the existence of a sequence $(\underset{\sim}{u}_j)$ of elements of $\mathscr{H}$ such that

$$\|\underset{\sim}{u}_j\| \le 1, \quad \lim_j (T\underset{\sim}{u}_j | \underset{\sim}{u}_j) = \lambda . \tag{2.4.10}$$

Since T is compact, passage to a suitable subsequence will arrange that $(T\underset{\sim}{u}_j)$ is convergent, say

$$\underset{\sim}{v} = \lim_j T\underset{\sim}{u}_j . \tag{2.4.11}$$

In addition to this, however, a subsequence $(\underset{\sim}{u}_{j_p})$ and $\underset{\sim}{u}_0 \in \mathscr{H}$ exist such that

$$\lim_p (\underset{\sim}{u}_{j_p} | \underset{\sim}{u}) = (\underset{\sim}{u}_0 | \underset{\sim}{u}) \tag{2.4.12}$$

for every $\underset{\sim}{u} \in \mathscr{H}$; see Exercise 2. 4. 8. The fact that (2. 4. 12) holds for every $\underset{\sim}{u} \in \mathscr{H}$ is expressed by saying that $\lim_p \underset{\sim}{u}_{j_p} = \underset{\sim}{u}_0$ <u>weakly in</u> $\mathscr{H}$. It follows easily from (2. 4. 10) and (2. 4. 12) that

$$\|\underset{\sim}{u}_0\| \le 1. \tag{2.4.13}$$

Also, for every $\underset{\sim}{u} \in \mathscr{H}$, (2. 4. 11) and (2. 4. 12) show that

$$\begin{aligned}(\underset{\sim}{v} | \underset{\sim}{u}) &= \lim_j (T\underset{\sim}{u}_j | \underset{\sim}{u}) = \lim_j (\underset{\sim}{u}_j | T^*\underset{\sim}{u}) \\ &= \lim_p (\underset{\sim}{u}_{j_p} | T^*\underset{\sim}{u}) = (\underset{\sim}{u}_0 | T^*\underset{\sim}{u}) = (T\underset{\sim}{u}_0 | \underset{\sim}{u}) .\end{aligned}$$

Hence

$$\underset{\sim}{v} = T\underset{\sim}{u}_0 . \tag{2.4.14}$$

Again,

$$(T\underset{\sim}{u}_{j_p} | \underset{\sim}{u}_{j_p}) = (\underset{\sim}{v} | \underset{\sim}{u}_{j_p}) + (T\underset{\sim}{u}_{j_p} - \underset{\sim}{v} | \underset{\sim}{u}_{j_p})$$

wherein the second summand on the right has an absolute value which, by (2. 4. 10), does not exceed $\|T\underset{\sim}{u}_j - \underset{\sim}{v}\|$. Reference to (2. 4. 10), (2. 4. 11) and (2. 4. 12) thus shows that

$$\lambda = \lim_p (T\underset{\sim}{u}_{j_p} | \underset{\sim}{u}_{j_p}) = \lim_p (\underset{\sim}{v} | \underset{\sim}{u}_{j_p}) = (\underset{\sim}{v} | \underset{\sim}{u}_0)$$

which, by (2.4.14), is equivalent to (2.4.8). Also, since $\lambda > 0$ ensures via (2.4.8) that $\underset{\sim}{u}_0 \neq 0$, the definition of λ shows that

$$\lambda \geq (T(\|\underset{\sim}{u}_0\|^{-1}\underset{\sim}{u}_0) | \|\underset{\sim}{u}_0\|^{-1}\underset{\sim}{u}_0) = \|\underset{\sim}{u}_0\|^{-2}(T\underset{\sim}{u}_0 | \underset{\sim}{u}_0) = \|\underset{\sim}{u}_0\|^{-2}\lambda ,$$

which implies that $\|\underset{\sim}{u}_0\| \geq 1$. This together with (2.4.13), yields (2.4.9).

To complete the proof of (a), choose any $\underset{\sim}{v} \in \mathscr{H}$ which is orthogonal to $\underset{\sim}{u}_0$ and has norm one. Then, for every complex α

$$\|\underset{\sim}{u}_0 + \alpha\underset{\sim}{v}\|^2 = 1 + |\alpha|^2 ,$$

and so the definition of λ shows that

$$(T(\underset{\sim}{u}_0 + \alpha\underset{\sim}{v}) | (\underset{\sim}{u}_0 + \alpha\underset{\sim}{v})) \leq \lambda(1 + |\alpha|^2) ,$$

that is

$$\bar{\alpha}(T\underset{\sim}{u}_0 | \underset{\sim}{v}) + \alpha(T\underset{\sim}{v} | \underset{\sim}{u}_0) \leq (\lambda - 1)|\alpha|^2 . \qquad (2.4.15)$$

Choose t real so that $k = e^{-it}(T\underset{\sim}{u}_0 | \underset{\sim}{v})$ is real and put $\alpha = re^{it}$ where r is real and non-zero. Then (2.4.15) yields

$$2rk \leq (\lambda - 1)r^2 . \qquad (2.4.16)$$

Assuming r to be positive, dividing by r and letting r tend to zero, (2.4.16) yields $k \leq 0$; similarly, assuming r to be negative and following the same procedure, it appears from (2.4.16) that $k \geq 0$. Hence $k = 0$ and so $(T\underset{\sim}{u}_0 | \underset{\sim}{v}) = 0$. Thus $T\underset{\sim}{u}_0$ is orthogonal to every $\underset{\sim}{v}$ which is orthogonal to $\underset{\sim}{u}_0$. Hence (see Exercise 2.4.9) $T\underset{\sim}{u}_0$ is a scalar multiple of $\underset{\sim}{u}_0$. Then (2.4.8) shows that this scalar multiple can be none other than $\lambda\underset{\sim}{u}_0$ and the proof of (a) is complete.

As to (b), it is trivial that $\dim M > 0$. On the other hand, compactness of T ensures that every bounded sequence extracted from M contains a convergent subsequence. This circumstance rules out the possibility that $\dim M = \infty$, for otherwise M would contain an infinite orthonormal sequence $(\underset{\sim}{e}_j)_{j=1}^{\infty}$, in which case $\|\underset{\sim}{e}_i - \underset{\sim}{e}_j\| = 2^{\frac{1}{2}}$ for every

$i \neq j$ and the bounded sequence $(\underset{\sim}{e}_j)_{j=1}^{\infty}$ of elements of M would contain no convergent subsequence.

2.4.5. Suppose the set of all coordinate functions specified in (CTi) of 2.4.1 to be indexed as a family $(u_\alpha)_{\alpha \in A}$. Then (CTi) asserts that to every $f \in L^2(G)$ and every $\varepsilon > 0$ corresponds a complex-valued function $\alpha \mapsto c_\alpha$ on A such that $\{\alpha \in A : c_\alpha \neq 0\}$ is finite and

$$\|f - \sum c_\alpha u_\alpha\|_2 \leq \varepsilon ,$$

the range of summation being (formally) A but in reality a certain finite subset of A. This is very far from asserting that to every $f \in L^2(G)$ corresponds a complex-valued function $\alpha \mapsto c'_\alpha$ on A such that the series $\sum c'_\alpha u_\alpha$ converges in some sense and has sum f. In spite of the gap, however, it turns out that there is a very reasonable sense in which the second assertion is true and implies the first assertion: this will appear in 2.7. A vital step in establishing this nice conclusion amounts to showing that there is subfamily $(u_\beta)_{\beta \in B}$, where B is a suitable subset of A, such that (i) every u_α $(\alpha \in A)$ is a linear combination of u_β's with $\beta \in B$; and (ii) the family $(u_\beta)_{\beta \in B}$ is orthogonal in $L^2(G)$. If this can be done, suitable choice of positive 'normalising factors' ν_β will arrange that $(\nu_\beta u_\beta)_{\beta \in B}$ is an orthonormal base in the Hilbert space $L^2(G)$ and elementary general theory will lead right to an expansion of the desired sort.

The next step is thus to prove orthogonality of a suitably large family of coordinate functions. This will be done in 2.6, the intermediary being a famous lemma about irreducible sets of endomorphisms (applying in particular to irreducible representations).

2.4.6. **Exercise.** Construct detailed proofs of 2.4.2(i) and (ii).

2.4.7. **Exercise.** Given that $f, g \in L^2(G)$ and $a \in G$, show that $f * L_a g(e) = g * f(a^{-1})$.

2.4.8. **Exercise.** Let $\mathscr{H}$ be a Hilbert space and $(\underset{\sim}{u}_j)$ a bounded sequence extracted from $\mathscr{H}$. Show that there exist a subsequence $(\underset{\sim}{u}_{j_p})$ and $\underset{\sim}{u}_0 \in \mathscr{H}$ such that (2.4.12) holds for every $\underset{\sim}{u} \in \mathscr{H}$. (Compare Edwards [3], Appendix B. 4.)

[Hint: Let H denote the closed subspace of generated by the $\underset{\sim}{u}_j$ and choose an orthonormal base $(\underset{\sim}{e}_r)_{r=1}^{\infty}$ for H. Use the diagonal process to prove the existence of a subsequence $(\underset{\sim}{u}_{j_p})$ such that $\alpha = \lim_p(\underset{\sim}{u}_{j_p} | \underset{\sim}{e}_r)$ exists for every r. Verify that

$$\sum_{r=1}^{\infty} |\alpha_r|^2 < \infty$$

and that $\underset{\sim}{u}_0 = \sum_{r=1}^{\infty} \alpha_r \underset{\sim}{e}_r$ satisfies all demands. This sketch deals with the case in which $\dim H = \infty$; if $\dim H < \infty$, a more elementary type of argument suffices.]

2.4.9. Let $\mathscr{H}$ be a Hilbert space, $\underset{\sim}{a}$ and $\underset{\sim}{b}$ elements of $\mathscr{H}$. Given that $\underset{\sim}{b}$ is orthogonal to every element of $\mathscr{H}$ which is orthogonal to $\underset{\sim}{a}$, show that $\underset{\sim}{b}$ is a scalar multiple of $\underset{\sim}{a}$.

2.4.10. Exercise. Use (2.3.4) and the uniqueness theorem to prove the associativity of convolution.

2.4.11. Exercise. Show that (UT_1) of 2.4.1 is a consequence of the following uniqueness assertion:

$$f \in C(G),\ \hat{f}(U) = 0 \text{ for every } U \in \hat{G} \Rightarrow f = \underline{0}.$$

2.5. Schur's lemma and its consequences

Recall that, if V is a linear space, End(V) denotes the set of all endomorphisms of V.

A subset Σ of End(V) is said to be irreducible if there exists no linear subspace of V, other than $\{\underset{\sim}{0}\}$ and V, which is invariant under Σ. (A subset M of V is invariant under Σ, if and only if $T(M) \subseteq M$ for every $T \in \Sigma$.)

If $\Sigma \subseteq \mathrm{End}(V)$ and $T \in \mathrm{End}(V)$, we shall write $T\Sigma$ (resp. ΣT) for $\{TS : S \in \Sigma\}$ (resp. $\{ST : S \in \Sigma\}$).

2.5.1. Schur's lemma. *Let V and V' be linear spaces and Σ and Σ' irreducible subsets of End(V) and End(V') respectively.*

Suppose that T is a linear mapping of V into V' such that

$$\Sigma' T = T\Sigma . \tag{2.5.1}$$

Then either (i) $T = 0$, or (ii) T is an isomorphism of V onto V'.

Proof. Consider $M' = T(V)$, which is a linear subspace of V'. Thanks to (2. 5. 1), M' is invariant under Σ' and so irreducibility of Σ' implies that M' is either $\{\underset{\sim}{0}\}$ or V'. Thus, either $T = 0$ or $T(V) = V'$.

By looking at $M = \ker T = \{\underset{\sim}{v} \in V : T\underset{\sim}{v} = \underset{\sim}{0}\}$, one sees similarly that either T is 0 or T is 1-1.

Hence, if $T \neq 0$, then T is both 1-1 and onto.

2. 5. 2. **Corollary.** Let V be a f. d. complex linear space and Σ an irreducible subset of End(V). If $T \in \mathrm{End}(V)$ is a commutator of Σ, i. e., if $T\Sigma = \Sigma T$, then T is a scalar multiple of the identity endomorphism of V.

Proof. Since V is f. d. and complex, there is a complex number λ such that $T_1 = T - \lambda I$ is not an isomorphism of V onto itself (I denoting the identity endomorphism of V). Now apply Schur's lemma, taking $V' = V$, $\Sigma' = \Sigma$ and replacing T by T_1. Since alternative (ii) is excluded by choice of λ, (i) must hold.

2. 5. 3. **Corollary.** Let V be a f. d. complex linear space. If End(V) contains a commutative irreducible subset, then $\dim V = 1$. In particular, any f. d. (complex) irreducible representation of an Abelian group is one-dimensional.

Proof. By 2. 5. 2, every element of the said subset is a scalar multiple of the identity; and then irreducibility implies that $\dim V = 1$.

2. 5. 4. **The Abelian case.** If G is Abelian, 2. 5. 3 shows that (as was heralded in 2. 2. 7 and 2. 3. 2) every $U \in \hat{G}$ is one-dimensional so that every character χ_U is a continuous multiplicative character of G. One may thus identify $\hat{G}$ with the set Γ of continuous multiplicative characters of G (i. e., the set of continuous homomorphisms of G into

T), the Fourier transform of $f \in L^1(G)$ being accordingly identified with the complex-valued function.

$$\chi \mapsto \hat{f}(\chi) = \int_G f(x)\overline{\chi(x)}dx$$

on Γ. (When Abelian groups alone are being studied, Γ is usually introduced directly and without any overt mention of representations.)

Under pointwise products, Γ becomes an Abelian group (infinite unless G is finite; see Exercise 2.6.9). If Γ is taken with the discrete topology, its bounded (automatically continuous) multiplicative characters turn out to be precisely those of the form $\chi \mapsto \chi(a)$ with $a \in G$. This statement is part of the famous Pontryagin duality theorem, a statement of which is worth including here.

The said duality theorem falls roughly into two parts:

(a) Start from any locally compact Abelian group G and form the group Γ of all bounded continuous multiplicative characters χ of G, the group operation in Γ being pointwise multiplication. (Having for the moment dropped the assumption that G be compact, the term 'bounded' must appear in the immediately preceding definition of Γ, boundedness being no longer a consequence of continuity. We might alternatively define Γ as the set of continuous homomorphisms of G into **T**, wherein the wording remains exactly as it did in the compact case.) Γ is then evidently an Abelian group. Topologise Γ by assigning to its neutral element a base of neighbourhoods comprising exactly the sets

$$W(K, \varepsilon) = \{\chi \in \Gamma : |\chi(x) - 1| \leq \varepsilon \quad \text{for every } x \in K\},$$

where K ranges over compact subsets of G and ε over all positive numbers. Then Γ proves to be a locally compact Abelian group termed the <u>dual</u> (or <u>character) group</u> of G.

(b) Now go through the same process, beginning with Γ instead of G. It is simple to check that each $a \in G$ generates a bounded continuous multiplicative character

$$\xi_a : \chi \mapsto \chi(a)$$

of Γ, i. e. an element of the dual of Γ. The crux of the duality law says that the mapping $a \mapsto \xi_a$ is an isomorphism of the topological group G onto the dual of Γ.

It also turns out that Γ is discrete if and only if G is compact.

For all of this, see Edwards [3], 2. 2. 1 and Hewitt and Ross [1], Chapter VI. Concerning duality for compact non-Abelian groups, see 2. 7. 7. below.

2. 5. 5. Exercise. Let $\mathcal{H}$ denote a 2-dimensional Hilbert space in which $(\underset{\sim}{e}_1, \underset{\sim}{e}_2)$ is an orthonormal base. Consider the representation U of the circle group **T** with representation space $\mathcal{H}$ defined by

$$U(e^{it}) : \begin{cases} \underset{\sim}{e}_1 \mapsto \cos t . \underset{\sim}{e}_1 + \sin t . \underset{\sim}{e}_2 \\ \underset{\sim}{e}_2 \mapsto -\sin t . \underset{\sim}{e}_1 + \cos t . \underset{\sim}{e}_2 \end{cases} .$$

By 2. 5. 3, U is reducible. Show how to reduce U into the sum of two irreducible representations of **T**.

2. 5. 7. Exercise. Show that G is Abelian if and only if $d(U) = 1$ for every $U \in \hat{G}$. (Assume (CTii) of 2. 4. 1.)

2. 6. The orthogonality relations

2. 6. 1. Let $U \in \hat{G}$ and let T be any endomorphism of $\mathcal{H}_U$. The endomorphism

$$T' = \int U(x)TU(x)^*dx$$

then commutes with every U(y). Since U is irreducible, Corollary 2. 5. 2 implies that T' is a scalar multiple of I_U. The value of this scalar is found immediately by taking the trace of both sides, remembering that $\operatorname{Tr} U(x)TU(x)^* = \operatorname{Tr} T$ since U(x) is unitary. Thus

$$\int U(x)TU(x)^*dx = d(U)^{-1} . \operatorname{Tr} T . I_U . \qquad (2. 6. 1)$$

A similar argument shows that

$$\int U(x)TV(x)^*dx = 0 \qquad (2. 6. 2)$$

if U, V $\in \hat{G}$, $U \neq V$ and T is any linear map: $\mathcal{H}_V \to \mathcal{H}_U$.

2.6.2. If orthonormal bases $(\underset{\sim}{u}_i)$ in $\mathcal{H}_U$ and $(\underset{\sim}{v}_j)$ in $\mathcal{H}_V$, are introduced, and if one writes

$$u_{ij}(x) = (\underset{\sim}{u}_i | U(x)\underset{\sim}{u}_j), \quad v_{hk}(x) = (\underset{\sim}{v}_h | V(x)\underset{\sim}{v}_k),$$

suitable choice of T in (2.6.1) and (2.6.2) lead to the orthogonality relations

$$\left.\begin{aligned} &\int u_{ij}(x)\overline{u_{hk}(x)}dx = \delta_{ih}.\delta_{jk}.d(U)^{-1} \\ &\qquad (1 \le i,\ j,\ k \le d(U)), \end{aligned}\right\} \qquad (2.6.3)$$

$$\left.\begin{aligned} &\int u_{ij}(x)\overline{v_{hk}(x)}dx = 0 \\ &(1 \le i,\ j \le d(U),\ \ 1 \le h,\ k \le d(V)). \end{aligned}\right\} \qquad (2.6.4)$$

2.6.3. If in (2.6.3) one puts $j = i$ and $h = k$ and sums over i and k, it appears that

$$\int \chi_U(x)\overline{\chi_U(x)}dx = 1. \qquad (2.6.5)$$

Likewise, from (2.6.4) there appears

$$\int \chi_U(x).\chi_V(x)dx = 0 \quad (U,\ V \in \hat{G},\ U \neq V). \qquad (2.6.6)$$

2.6.4. Actually (2.6.2) yields more than this: if one replaces T by FTG*, where F and G are arbitrary endomorphisms of $\mathcal{H}_U$ and $\mathcal{H}_V$ respectively, there results the formula

$$\int U(x)FTG^*V(x)^*dx = 0;$$

and from this may be deduced by rather tedious calculation the equivalent formulae

$$\left.\begin{aligned} &\int Tr[U(x)F].\overline{Tr[GV(x)]}dx = 0 \\ &\int Tr[FU(x)^*].\overline{Tr[GV(x)^*]}dx = 0 \end{aligned}\right\} \qquad (2.6.7)$$

and in particular (see (2.3.6))

$$\int f * \chi_U(x).\overline{g * \chi_V(x)}dx = 0 . \tag{2.6.8}$$

Similarly, (2.6.1) gives

$$\int U(x)FTG^*U(x)^*dx = d(U)^{-1}.\, Tr[FTG^*].\, I_U$$

and thence

$$\left.\begin{aligned} &\int Tr[U(x)F].\,\overline{Tr[U(x)G]}dx = d(U)^{-1}.\, Tr[FG^*] \\ &\int Tr[FU(x)^*].\,\overline{Tr[GU(x)^*]}dx = d(U)^{-1}.\, Tr[FG^*] \end{aligned}\right\} \tag{2.6.9}$$

and in particular (see (2.3.6))

$$\int f * \chi_U(x).\overline{g * \chi_U(x)}dx = d(U)^{-1}.\, Tr[\hat{f}(U)\hat{g}(U)^*] . \tag{2.6.10}$$

Some indications of the intermediate calculations are given in Appendix B.1, where it is shown too that the Fourier transform of the function

$$h(x) = d(U).\, Tr[HU(x)^*] , \tag{2.6.11}$$

H being any endomorphism of $\mathscr{H}_U$, is given by

$$\hat{h}(V) = \begin{cases} H & \text{if } V = U \\ 0 & \text{if } V \neq U, \quad V \in \hat{G} \end{cases} \tag{2.6.12}$$

2.6.5. **Remarks.** The preceding arguments really show that, if U and V are arbitrary continuous irreducible unitary representations of G (not necessarily elements of the chosen set $\hat{G}$ of representatives), then $\int \chi_U \overline{\chi}_V \, d\mu$ is 1 or 0 according as U and V are or are not equivalent.

This means in particular that one could very well use the set Γ of all characters of elements of $\hat{G}$ to index $\hat{G}$, i.e., that there is a bijection $\chi \mapsto U_\chi$ of Γ onto $\hat{G}$, U_χ denoting the unique element of $\hat{G}$ whose character is χ. This procedure will not be adopted in these notes, though it is used by some authors.

The advantages of this procedure are not fully apparent until one has picked out or adequately described the elements of Γ in a fashion making no explicit reference to representations. In case G is Abelian, this is easy: the elements of Γ are precisely the continuous homomorphisms of G into the circle group $\mathbf{T}$, i. e., precisely the continuous multiplicative characters of G (see 2. 5. 4). If G is not assumed to be Abelian, the desired sort of description is not so obvious. However, there are various ways of solving the problem, two of which follow by way of illustration.

(i) Γ consists of those $f \in C(G)$ which satisfy $\|f\|_2 = 1$ and which, for some (f-dependent) complex number $c \neq 0$, are such that

$$\int f(xax^{-1}b)dx = c^{-1}f(a)f(b)$$

for every $a, b \in G$.

A proof of this appears in Loomis [1], 39C; see also Hewitt and Ross [1], (27. 53). (Loomis' proof is expressed in terms rather different from those used in these notes; all the same, an interested reader should not experience much difficulty in translation.)

(ii) Denote by K the set of non-zero central functions $k \in C(G)$ such that $k * k = ck$ for some (k-dependent) positive number c. Say that $k \in K$ is indecomposable if and only if it is not expressible in the form $k_1 + k_2$ with $k_1, k_2 \in K$, unless k_1 is a constant multiple of k. Then Γ comprises precisely those elements k of K which are indecomposable and satisfy $\|k\|_2 = 1$.

For a proof of this, see Exercise 2. 6. 8.

In spite of these possibilities, the procedure will not be adopted in these notes. We shall make systematic use of Γ only in case G Abelian, and then Γ is to be understood in the manner described in 2. 3. 3 and 2. 5. 4, i. e., Γ is to denote the dual group of continuous multiplicative characters of G. In other cases we continue to use the set $\hat{G}$ of representations.

2. 6. 6. The Abelian case. In view of 2. 5. 4, the orthogonality relations now take the form of the assertion: if $\chi_1, \chi_2 \in \Gamma$, then

$$\int \chi_1 \overline{\chi}_2 d\mu = \begin{cases} 1 \text{ if } \chi_1 = \chi_2 \\ 0 \text{ otherwise.} \end{cases} \tag{2.6.13}$$

What is more, this can be proved very simply indeed and without any dealings with representations in general. Thus, if $\chi \in \Gamma$ and $a \in G$, then $L_a \chi = \overline{\chi(a)}\chi$ and so invariance of the integral gives

$$\int \chi d\mu = \int L_a \chi d\mu = \chi(a) \int \chi d\mu .$$

If $\chi \neq \underline{1}$, a can be chosen so that $\chi(a) \neq 1$, in which case the last formula shows that $\int \chi d\mu = 0$. This, combined with normalisation of the invariant integral, leads to (2.6.13) when χ is taken to be $\chi_1 \overline{\chi}_2 = \chi_1 \chi_2^{-1}$.

2.6.7. Fourier characterisation of central functions. Let $k \in L^1(G)$. In order that k be central (see 2.1.6), it is necessary and sufficient that, for every $U \in \hat{G}$, $\hat{k}(U)$ be a scalar multiple of I_U.

Proof. As for necessity, if k is central, (2.1.8) and (2.3.4) show that, for every $U \in \hat{G}$, $\hat{k}(U)\hat{f}(U) = \hat{f}(U)\hat{k}(U)$ for every $f \in L^1(G)$. By (2.6.11) and (2.6.12), this implies that, for every $U \in \hat{G}$, $\hat{k}(U)$ commutes with every element of $\mathrm{End}(\mathscr{H}_U)$ and so, by 2.5.2, is a scalar multiple of I_U.

Turning to sufficiency, if $\hat{k}(U)$ is, for every $U \in \hat{G}$, a scalar multiple of I_U, (2.3.4) shows that for every $f \in L^1(G)$ the functions $f * k$ and $k * f$ have the same Fourier transform. That k is central, now follows from the uniqueness theorem (UT_1) of 2.4.1.

2.6.8. Exercise. Prove 2.6.5(ii).

2.6.9. Exercise. Show that G is finite if and only if $\hat{G}$ is finite.

2.6.10. Exercise. Let G be a compact group. For $f \in C(G)$ and n a positive integer, define $C_n f$ by recurrence so that $C_1 f = f$ and $C_{n+1} f = f * C_n f$. Prove that G is Abelian if and only if $\underline{0}$ $(= \underline{0}_G)$ is the only element f of C(G) satisfying $C_n f = \underline{0}$ for some positive integer n.

2.6.11. Exercise. Let G be a compact Abelian group with dual group Γ. Assume (what will be established in 2.8.8) that the linear combinations of elements of Γ are dense in $C(G)$. Use the orthogonality relations to verify the following two statements.

(i) If $(G_n)_{n=1}^{\infty}$ is an increasing sequence of closed subgroups of G whose union is dense in G, then

$$\int f d\mu_G = \lim_{n\to\infty} \int f d\mu_{G_n} \qquad (2.6.14)$$

for every $f \in C(G)$.

(ii) If $a \in G$ generates a dense subgroup of G, then

$$\int f d\mu_G = \lim_{n\to\infty} (2n)^{-1} \sum_{|m| \le n} f(a^m) \qquad (2.6.15)$$

for every $f \in C(G)$.

Illustrate by examples.

Remarks. Both formulae (2.6.14) and (2.6.15) continue to hold for certain discontinuous complex-valued functions f on G; cf. Exwards [3], Exercise 2.15.

2.7. Fourier series in $L^2(G)$

2.7.1. Fourier series. The orthogonality relations (2.6.3) and (2.6.4) show that the family

$$(d(U)^{\frac{1}{2}} u_{ij})_{1 \le i, j \le d(U)}, \quad U \in \hat{G}$$

is orthonormal in $L^2(G)$, and the completeness theorem (CTi) in 2.4.1 shows that it is an orthonormal base in $L^2(G)$. The associated orthogonal expansion of $f \in L^2(G)$ is the <u>Fourier series of</u> f, namely

$$\Sigma d(U) \sum_i \sum_j \{ \int f(y) \overline{u_{ij}(y)} dy \} u_{ij} ,$$

Herein the term arising from the representation U is, apart from the factor $d(U)$,

$$\sum_i \sum_j \int f(y)\overline{u_{ij}(y)}dy.\, u_{ij}(x)\,,$$

which is none other than

$$f * \chi_U(x) = Tr[\hat{f}(U)U(x)^*] = \chi_U * f(x)\,.$$

The Parseval formula will read

$$\int |f(x)|^2 dx = \Sigma d(U) \sum_i \sum_j \left| \int f(y)\overline{u_{ij}(y)}dy \right|^2$$

and the 'U-term' in this is precisely

$$d(U).\, Tr[\hat{f}(U)\hat{f}(U)^*]\,.$$

For more details, see Appendix B. 2.

2. 7. 2. Convergence in $L^2(G)$. A rigorous argument proceeds along customary lines, using the orthogonality relations. Thus, suppose $f \in L^2(G)$. If P denotes a finite subset of $\hat{G}$ and

$$f_P(x) = \sum_{U \in P} d(U).\, Tr[\hat{f}(U)U(x)^*]\,,$$

(2. 6. 7) and (2. 6. 9) give

$$\int |f(x) - f_P(x)|^2 dx = \int \cdot |f(x)|^2 dx - \sum_{U \in P} d(U) Tr[\hat{f}(U)\hat{f}(U)^*]\,,$$

whence follows Bessel's inequality:

$$\Sigma d(U).\, Tr[\hat{f}(U)\hat{f}(U)^*] \le \int |f(x)|^2 dx\,. \qquad (2.7.1)$$

On the other hand, if the function Φ assigns to each $U \in \hat{G}$ an endomorphism of $\mathscr{H}_U$ in such a way that $\Sigma d(U).\, Tr[\Phi(U)\Phi(U)^*] < \infty$, and if

$$\phi_P(x) = \sum_{U \in P} d(U).\, Tr[\Phi(U)U(x)^*]$$

for every finite subset P of $\hat{G}$, then the orthogonality relations show that

$$\int |\phi_Q(x) - \phi_P(x)|^2 dx = \sum_{U \in Q \setminus P} d(U).\, Tr[\Phi(U)\Phi(U)^*]$$

for finite sets $Q \supset P$. Hence the family (ϕ_P) is Cauchy, when the P's are directed by inclusion. Since $L^2(G)$ is complete, $\phi = \lim_P \phi_P$ exists in the L^2-sense. Since the Fourier transformation is linear and continuous $(\|\hat{f}(U)\| \le \|f\|_1 \le \|f\|_2)$, (2. 6. 11) and (2. 6. 12) show that $\hat{\phi} = \Phi$ and

$$\int |\phi(x)|^2 dx = \lim_P \int |\phi_P(x)|^2 dx = \Sigma d(U). \operatorname{Tr}[\hat{\phi}(U)\hat{\phi}(U)^*] ,$$

the sum extending over all $U \in \hat{G}$.

2. 7. 3. The Parseval formulae. We are at liberty to take $\Phi = \hat{f}$, whenever $f \in L^2(G)$. Then the completeness theorem (UT_2) of 2. 4. 1 tells us that the ϕ obtained in 2. 7. 2 is equal a. e. to f. Thus, writing $\Sigma \ldots$ in place of $\Sigma_{U \in \hat{G}} \ldots$,

$$f(x) = \Sigma d(U). \operatorname{Tr}[\hat{f}(U)U(x)^*] = \Sigma d(U). f * \chi_U(x) , \qquad (2. 7. 2)$$

the series being the strong limit in $L^2(G)$ of its finite partial sums; moreover (2. 7. 1) is sharpened into the Parseval formula

$$\int |f(x)|^2 dx = \Sigma d(U). \operatorname{Tr}[\hat{f}(U)\hat{f}(U)^*] . \qquad (2. 7. 3)$$

More generally, if f and g are in $L^2(G)$, then

$$\int f(x)\overline{g(x)} dx = \Sigma d(U). \operatorname{Tr}[\hat{f}(U)\hat{g}(U)^*] , \qquad (2. 7. 4)$$

the series on the right converging absolutely; this is the so-called polarised (form of the) Parseval formula.

Note that (2. 7. 4) can also be written

$$f * \tilde{g}(e) = \Sigma d(U). \operatorname{Tr}[\hat{f}(U)\hat{g}(U)^*] \qquad (2. 7. 5)$$

for f, g in $L^2(G)$. Consequently, if $\phi \in L^1(G)$, then

$$f * \phi * \tilde{g}(e) = \Sigma d(U). \operatorname{Tr}[\hat{f}(U)\hat{\phi}(U)\hat{g}(U)^*] , \qquad (2. 7. 6)$$

as follows from (2. 3. 3), (2. 3. 4) and (2. 7. 5). Equation (2. 7. 6) holds for all f, g in $L^2(G)$ and the series converges absolutely.

We shall return in 2. 9 to consideration of the pointwise convergence of the Fourier series (2. 7. 2) for restricted classes of functions f.

Remark. The substance of 2. 7. 2 and 2. 7. 3 might be summarised by saying that $L^2(G)$ is the internal Hilbertian direct sum (see Exercise 2. 2. 16) of its f. d. subspaces $\chi_U * L^2(G)$, one for each $U \in \hat{G}$. Similar and more general results of this sort will be studied in 2. 12 below.

2. 7. 4. The Riemann-Lebesgue lemma. *If* $f \in L^1(G)$, *then*

$$\lim_{U \in \hat{G},\, U \to \infty} \|\hat{f}(U)\| = 0 . \tag{2. 7. 7}$$

(The interpretation of (2. 7. 7) is explained in Exercise 1. 2. 5.)

Proof. From Appendix A, formula (A. 2. 8) one has the estimate

$$\|A\| \le (\operatorname{Tr} AA^*)^{\frac{1}{2}} \le d^{\frac{1}{2}} \|A\| \tag{2. 7. 8}$$

for any $A \in \operatorname{End}(\mathscr{H}_U)$, where $\|A\|$ denotes the usual operator norm of A and d denotes the dimension $d(U)$ of $\mathscr{H}_U$. The conclusion (2. 7. 7) follows at once from (2. 7. 3) and (2. 7. 8), provided $f \in L^2(G)$.

Now suppose that $f \in L^1(G)$ and $\varepsilon > 0$. Choose $f_1 \in L^2(G)$ so that $\|f - f_1\|_1 \le \frac{1}{2}\varepsilon$. By what has just been established,

$$F = \{U \in \hat{G} : \|\hat{f}_1(U)\| \ge \tfrac{1}{2}\varepsilon\}$$

is finite. On the other hand, in view of (2. 3. 2) and the choice of f_1,

$$\{U \in \hat{G} : \|\hat{f}(U)\| \ge \varepsilon\} \subseteq F ,$$

which implies (2. 7. 7).

Remark. Even if G is Abelian, there are senses in which (2. 7. 7) is the most one can say... and that even for continuous f. For example, supposing G to be infinite Abelian and $\chi \mapsto \varepsilon(\chi)$ to be any preassigned non-negative function on Γ such that $\lim_{\chi \in \Gamma,\, \chi \to \infty} \varepsilon(\chi) = 0$, there exist functions $f \in C(G)$ such that the formula $\hat{f}(\chi) = \underline{0}(\varepsilon(\chi))$ is false. To see this, note that one may choose $\chi_n \in \Gamma$ $(n = 1, 2, \ldots)$ so that $\Sigma_n\, n\, \varepsilon(\chi_n) < \infty$; then

$$f = \Sigma_n\, n\, \varepsilon(\chi_n) \chi_n \in C(G)$$

and yet $\hat{f}(\chi_n) = n\,\varepsilon(\chi_n)$ for every n (as a consequence of the orthogonality relations (2.6.13)).

For a discussion of the (necessarily more complicated) non-Abelian case, see Mayer [1].

2.7.5. Trigonometric polynomials. The substance of 2.7.2 suggests that considerable importance be attached to the set of continuous functions f such that

$$\{U \in \hat{G} : \hat{f}(U) \neq 0\}$$

is finite. Such functions are termed trigonometric polynomials (t.p.s for short) on G. (The name is a take-over from the special case in which $G = T$: see 2.1.15, 2.5.4 and 2.7.6.) The symbol $T(G)$ will denote the set of all t.p.s on G.

As follows from the completeness theorem and the orthogonality relations, the t.p.s on G are just the functions of the form

$$x \mapsto \sum_{j=1}^{n} c_j \mathrm{Tr}[A_j U_j(x)^*],$$

where the c_j are complex numbers, the $U_j \in \hat{G}$ and $A_j \in \mathrm{End}(\mathscr{H}_{U_j})$ for each j. In other words, $T(G)$ is just the linear subspace of $C(G)$ generated by all coordinate functions associated with elements of $\hat{G}$.

Statement 2.4.1(i) says exactly that $T(G)$ is dense in $L^2(G)$. As will appear in 2.8.8 and 2.9.3, $T(G)$ is dense in $C(G)$ and in $L^p(G)$ whenever $1 \leq p < \infty$: in fact, any one statement of this type might be thought of as a variant of the completeness theorem.

Independently of the characterisation of $T(G)$ in terms of coordinate functions, the results of 2.3.1 show at once that $T(G)$ is bi-invariant (under translations), that $T(G)^{\sim} \subseteq T(G)$, and that $M(G) * T(G) * M(G) \subseteq T(G)$. The last formula (which, of course, signifies that $\alpha * t * \beta$ is a t.p. whenever t is a t.p. and α and β are Radon measures on G) implies that $L^p(G) * T(G)$, $T(G) * L^p(G)$ and $L^p(G) * T(G) * L^p(G)$ are all subsets of $T(G)$. Thus, if one regards $L^p(G)$ as an algebra in which the product is convolution (see Edwards [3], 3.1.7), then $T(G)$ is at once a left ideal, a right ideal and a two-sided ideal in $L^p(G)$; see 2.12.1 below.

There remain two other crucial properties of T(G) which are not so evident, namely:

(i) $T(G)^{-} \subseteq T(G)$;

(ii) $T(G).T(G) \subseteq T(G)$.

The first says that T(G) is invariant under complex conjugation; the second that the pointwise product of two t. p. s is a t. p. (i. e., that T(G), as well as being a linear subspace of C(G), is also a subalgebra thereof relative to pointwise operations).

Sketch proof of (i). It suffices to prove this: if $U \in \hat{G}$ and $t : x \mapsto (\underset{\sim}{u}|U(x)\underset{\sim}{v})$ is a coordinate function associated with U, then $\overline{t}$ is a coordinate function associated with some $V \in \hat{G}$. To this end, choose a conjugate linear map C of $\mathscr{H}_U$ onto itself such that $(C\underset{\sim}{a}|C\underset{\sim}{b}) = (\underset{\sim}{b}|\underset{\sim}{a})$ for every $\underset{\sim}{a}, \underset{\sim}{b} \in \mathscr{H}_U$ and $C^2 = I_U$ ... for example, choose any orthonormal base $(\underset{\sim}{e}_i)$ for $\mathscr{H}_U$ and define $C : \Sigma_i \alpha_i \underset{\sim}{e}_i \mapsto \Sigma_i \overline{\alpha}_i \underset{\sim}{e}_i$. It is easy to see that $V_0 : x \mapsto CU(x)C$ is a representation of G which is continuous, unitary and irreducible (since U has those properties). Also, for every $x \in G$,

$$t(x) = (\underset{\sim}{u}|U(x)\underset{\sim}{v}) = (CU(x)\underset{\sim}{v}|C\underset{\sim}{u}) = \overline{(C\underset{\sim}{u}|V_0(x)C\underset{\sim}{v})}$$
$$= \overline{s(x)} \text{ , say,}$$

s being a coordinate function associated with V_0. By 2.2.8(e), V_0 is unitarily equivalent to some $V \in \hat{G}$, say $V_0(x) = W^{-1}V(x)W$ for every $x \in G$ and some fixed unitary W. Then

$$s(x) = (C\underset{\sim}{u}|V_0(x)C\underset{\sim}{v}) = (C\underset{\sim}{u}|W^{-1}V(x)WC\underset{\sim}{v})$$
$$= (\underset{\sim}{u}'|V(x)\underset{\sim}{v}') \text{ ,}$$

where $\underset{\sim}{u}' = WC\underset{\sim}{u}$, $\underset{\sim}{v}' = WC\underset{\sim}{v}$ belong to $\mathscr{H}_V$, showing that s is a coordinate function associated with $V \in \hat{G}$. Since $\overline{t} = s$, (i) is established.

Sketch proof of (ii). It suffices to show that, if U' and U" are elements of $\hat{G}$, and if

$$t' : x \mapsto (\underset{\sim}{u}'|U'(x)\underset{\sim}{v}') \text{ and } t'' : x \mapsto (\underset{\sim}{u}''|U''(x)\underset{\sim}{v})$$

are associated coordinate functions, then $t = t't''$ is a t. p. To this end, let $\mathscr{H}'$ and $\mathscr{H}''$ be the representation spaces of U' and U'' respectively. Form (see Halmos [2], Sections 25 and 52) the tensor product $\mathscr{H} = \mathscr{H}' \otimes \mathscr{H}''$ and $U : x \mapsto U(x) = U'(x) \otimes U''(x)$. Then U is a continuous unitary representation of G and

$$t(x) = (\underset{\sim}{u} | U(x)\underset{\sim}{v})$$

for every $x \in G$, where $\underset{\sim}{u} = \underset{\sim}{u}' \otimes \underset{\sim}{u}''$ and $\underset{\sim}{v} = \underset{\sim}{v}' \otimes \underset{\sim}{v}''$. In general, U is not irreducible. However, it may (see 2. 2. 5) be decomposed into a finite sum of irreducible continuous unitary representations V_i, so that t appears as a finite sum $\Sigma_i t_i$, where t_i is a coordinate function associated with V_i. By 2. 2. 8(e) once again, each V_i is unitarily equivalent to some $U_i \in \hat{G}$ and then (as in (i) above) t_i is seen to be a coordinate function associated with U_i.

2. 7. 6. The Abelian case. Here the Fourier series of f reduces to the form

$$\Sigma_{\chi \in \Gamma} \hat{f}(\chi)\chi , \qquad (2. 7. 9)$$

a sum extended over the elements of the dual group Γ described in 2. 5. 4. The Parseval formulae read

$$\left.\begin{aligned} \int_G |f(x)|^2 dx &= \Sigma_{\chi \in \Gamma} |\hat{f}(\chi)|^2 , \\ \int_G f(x)\overline{g(x)} dx &= \Sigma_{\chi \in \Gamma} \hat{f}(\chi)\overline{\hat{g}(\chi)} \end{aligned}\right\} \qquad (2. 7. 10)$$

for $f, g \in L^2(G)$. The series on the right in (2. 7. 10) are absolutely convergent whenever $f, g \in L^2(G)$; but the convergence of the series on the right in (2. 7. 9) is much more delicate (even if f is continuous); see 2. 9. 7 below.

The Riemann-Lebesgue lemma asserts that $\hat{f}(\chi) \to 0$ as $\chi \to \infty$ for every $f \in L^1(G)$.

$T(G)$ now consists exactly of all finite sums

$$\Sigma_j c_j \chi_j ,$$

where the c_j are complex numbers and the χ_j are elements of Γ. In particular, if $G = \mathbf{T}$ is the circle group, the t. p. s on G are just the functions

$$e^{it} \mapsto \Sigma_j c_j e^{ijt} \quad \text{(finite sum)} :$$

see 2. 2. 15. The properties of $T(G)$ proved in 2. 7. 5(i) and (ii) are much more evident in the present (Abelian) case.

2. 7. 7. Non-Abelian duality: Tannaka's theorem. In 2. 5. 4 there appears a brief description of the Pontryagin duality theorem according to which, if G is compact (or even merely locally compact) and Abelian, G is recoverable from the dual group Γ, the group structure of the latter being expressible in terms of pointwise products and complex conjugation. (If χ and χ' are elements of Γ, their group product $\chi\chi'$ is just their pointwise product qua complex-valued functions on G, and the group inverse χ^{-1} is just the complex conjugate function $\overline{\chi}$.) It is natural to ask what analogue, if any, of all this is applicable when G is compact non-Abelian? What is to be used as a suitable 'dual object' in place of Γ?

The fact, noted in 2. 2. 9, that $\hat{G}$ (unlike Γ in the Abelian case) is not intrinsically related to G, suggests that it is perhaps not the sought-for dual object. That it certainly is not, can be shown by examples (Hewitt and Ross [1], (27. 62. f)) of non-isomorphic G's having isomorphic $\hat{G}$'s.

A clue to the right course is obtainable by looking again at the compact Abelian case, where it is known that G is recoverable from Γ, and viewing Γ in a fresh way. Specifically, the trick is to view Γ as a subset of $T(G)$ and to recall that the group structure of Γ is inherited from pointwise product and complex conjugation applied to $T(G)$. Since, furthermore, G is realisable as the set of bounded multiplicative characters ξ of Γ (see 2. 5. 4(b)), it seems that one should look for the functions on $T(G)$ which are, in some sense, the extensions to $T(G)$ of the ξ's on $\Gamma \subseteq T(G)$. Now it is easy to check that each ξ can be extended into a function M on $T(G)$ in the following rather natural way: each $t \in T(G)$ can be expressed uniquely as a finite sum

$$t = \Sigma_j c_j \chi_j ,$$

where $j \mapsto \chi_j$ is an injection into Γ and the c_j are complex numbers; $M(t)$ is accordingly defined to be $\Sigma_j c_j \xi(\chi_j)$. This definition of M evidently makes it a non-zero multiplicative linear functional on $T(G)$ such that

$$M(\bar{t}) = \overline{M(t)} \qquad (2.7.11)$$

for every $t \in T(G)$. Denote by $\mathscr{G}$ the set of all such functionals on $T(G)$. It is equally evident that, if $M \in \mathscr{G}$, then $\xi = M|\Gamma$ is a bounded multiplicative character of Γ. Moreover, if $a \in G$, the associated 'evaluation functional'

$$E_a : t \mapsto t(a)$$

is plainly an element of $\mathscr{G}$. The essence of the Pontryagin duality theorem stated in 2.5.4 may (in the compact Abelian case at any rate) now be expressed in the form

$$\mathscr{G} = \{E_a : a \in G\} . \qquad (2.7.12)$$

(It is important to indicate here that, although the Pontryagin duality theorem applies to all locally compact Abelian G, that theorem is not properly rendered by the assertion (2.7.12) unless G is compact. The reason is that, if G is non-compact, Γ is non-discrete and one has then to distinguish between those bounded multiplicative characters of Γ which are continuous and those which are not. See 2.7.8 below.)

Now suppose that G is compact but not necessarily Abelian. Then $\mathscr{G}$ can be defined exactly as above, but the assertion (2.7.12) is no longer part of the Pontryagin duality theorem: instead it is part of a new duality theorem due to Tannaka. The proof of (2.7.12), although it is by no means the end of the road (one has thereafter to form $\mathscr{G}$ into a compact group and prove this to be isomorphic to G), is a vital step in establishing the Tannaka duality theorem. For details concerning all this, see Hewitt and Ross [1], Section 30.

2.7.8. As has been said, (2.7.12) does not faithfully render the Pontryagin duality theorem when G is locally compact Abelian and non-compact: in fact, (2.7.12) is generally false for such groups G. This is because Γ is non-discrete and in general there will exist discontinuous bounded multiplicative characters ξ of **Γ**, i.e., bounded multiplicative characters ξ of Γ which are not of the form $\xi(\chi) = \chi(a)$ with $a \in G$. The case G = R (see Exercise 2.2.14) illustrates the possibilities.

2.7.9. **Exercise.** Let $f \in L^2(G)$. Show that there exists a central function $k \in L^2(G)$ such that $\|f - k\|_2$ is a minimum, and that this

$$k = \Sigma_{U \in \hat{G}} (\mathrm{Tr}\, \hat{f}(U)) \chi_U ,$$

the series converging in $L^2(G)$. Deduce that

$$\Sigma_{U \in \hat{G}} |\mathrm{Tr}\, \hat{f}(U)|^2 \leq \|f\|_2^2 ,$$

equality holding if and only if f is central.

2.7.10. **Exercise.** Let G be a compact Abelian group with dual group Γ. Write Hom(**R**, G) for the set of all continuous group homomorphisms of **R** into G. It is known (Riss [1], p. 52, Proposition 3; Hewitt and Ross [1], (25.20)) that, if G is connected, then $G_0 = \{h(1) : h \in \mathrm{Hom}(\mathbf{R}, G)\}$ is everywhere dense in G; assume this throughout the present exercise.

Show that there exists a function $l : \Gamma \times \mathrm{Hom}(\mathbf{R}, G) \to \mathbf{R}$ such that

$$\chi(h(r)) = \exp\{i l(\chi, h) r\} \qquad (2.7.13)$$

for every $(\chi, h, r) \in \Gamma \times \mathrm{Hom}(\mathbf{R}, G) \times \mathbf{R}$.

Denote by $\mathscr{A}(G)$ the set of $f \in C(G)$ with the property: to every $h \in \mathrm{Hom}(\mathbf{R}, G)$ corresponds a number $\varepsilon = \varepsilon(f, h) > 0$ such that

$$\Sigma_{\chi \in \Gamma} \exp\{\varepsilon |l(\chi, h)|\}. |\hat{f}(\chi)| < \infty . \qquad (2.7.14)$$

Prove that, if G is connected, and if $f \in \mathscr{A}(G)$ vanishes on

some non-void open subset W of G, then $f = \underline{0}$.
[Hints: Show that one may assume without loss of generality that W is a neighbourhood of the identity element e of G. Consider functions of the form $f \circ h$, where $h \in \mathrm{Hom}(\mathbf{R}, G)$, showing that each of these is the restriction to $\mathbf{R}$ of a function analytic in a strip $\{r \in \mathbf{C} : |\mathrm{Im}\, r| < \varepsilon\}$; see (2.7.13) and (2.7.14).]

Remarks. (i) It is easy to verify that $\mathscr{A}(G)$ is an algebra under pointwise operations, and that $\mathscr{A}(G) * M(G) \subseteq \mathscr{A}(G)$. The above results show that $\mathscr{A}(G)$ behaves in at least one important respect like an algebra of analytic functions (even though G will bear little resemblance to an analytic manifold; see Edwards [5], 2.5). In pushing these ideas further, it would be convenient to use the fact that connectedness of G is necessary and sufficient that Γ may be ordered (see Hewitt and Ross [1], (24.25) and Rudin [1], 8.1.2). This feature is the basis for the usual approach to the study of functions on G having Fourier series of so-called 'analytic type', which forms an important section of commutative harmonic analysis; see Rudin [1], Chapter 8.

(ii) Conditions similar to, but weaker than, (2.7.14) lead to quasi-analyticity, and thence to similar conclusions; see Edwards [3], Exercise 2.8 and the references cited there.

(iii) The arguments can be extended to the case in which G is merely locally compact Abelian and connected. However, in this case it is more natural to start with a locally compact Abelian group Γ which is torsion-free, and regard Γ as the dual of a suitable locally compact connected Abelian group G (recall the Pontryagin duality law mentioned in 2.5.4 above and Hewitt and Ross [1], (25.24) once more). The result then takes the following form. Suppose m is a complex Radon measure on Γ with the property that to each $h \in \mathrm{Hom}(\mathbf{R}, G)$ corresponds a number $\varepsilon = \varepsilon(m, h) > 0$ such that

$$\int_{\Gamma}^{*} \exp\{\varepsilon\, |l(\chi, h)|\}\, d|m|(\chi) < \infty, \tag{2.7.15}$$

which is the 'continuous' analogue of (2.7.14). (In (2.7.15), $|m|$ denotes the non-negative Radon measure on Γ equal to $m_1 + m_2 + m_3 + m_4$, where $m = m' + im''$, m' and m'' being real Radon measures on Γ, and $m' = m_1 - m_2$, $m'' = m_3 - m_4$ are the minimal decompositions of m' and m'' into differences of two non-negative Radon measures on Γ, as in Exercise 1. 2. 6.) Consider the transform

$$m^* : x \in G \mapsto \int_\Gamma \chi(x) dm(\chi) .$$

The conclusion is that, if m is not the zero measure, then m^* is non-vanishing at some point of every non-void open subset W of G.

2. 8. Positive definite functions

2. 8. 0. Introductory remarks. A section devoted to positive definite functions seems justifiable for at least two reasons.

Firstly, as was mentioned in 2. 0. 2(iv), it is possible to make positive definite functions, rather than unitary representations, the basic tools; this procedure, due to Godement, Gelfand and Raikov, is at least partially successful for general locally compact groups. What follows in this section helps to clarify the close connections between unitary representations and positive definite functions, thereby rendering plausible the possibility of using the latter as the fundamental objects; see 2. 8. 10(ii).

Secondly, as Bochner showed in the case of $G = \mathbf{R}$, the use from the outset of positive definite functions provides an effective approach to relatively concrete problems concerning Fourier integrals (there being no necessary overt reference to representations or to abstract versions of harmonic analysis). This sort of application will be illustrated in 2. 9.

2. 8. 1. Definition. A complex-valued function ϕ on G is said to be <u>positive definite</u> (PD for short) if and only if $\phi \in L^1(G)$ and

$$f * \phi * \tilde{f}(e) = \int\int \phi(y^{-1}x)\bar{f}(x)f(y)dxdy \geq 0 \qquad (2.8.1)$$

for every $f \in C(G)$.

The formula (2. 8. 1) serves equally well to define positive definite measures, but little or no use will be made of this concept.

The reader should verify that, whenever $\psi \in L^1(G)$, both $\psi * \tilde{\psi}$ and $\tilde{\psi} * \psi$ are PD. (A quick proof of this will follow from 2. 8. 2(i), (2. 3. 3) and (2. 3. 4).)

On using the extended versions of the formulae appearing in Exercise 2. 1. 15, it may be shown that $\overline{\phi}$ and $\check{\phi}$ (and hence also $\tilde{\phi}$, for which case see alternatively (2. 3. 3) and 2. 8. 2(i)) are PD whenever ϕ is PD.

As the next result shows, PD functions may be characterised quite simply in terms of their Fourier transforms.

2. 8. 2. Theorem. (i) A function $\phi \in L^1(G)$ is PD if and only if $\hat{\phi}(U)$ is p. s. a. (= positive self-adjoint; see Appendix A. 0. 1) for every $U \in \hat{G}$.

(ii) If ϕ is PD, then $\tilde{\phi} = \phi$ as elements of $L^1(G)$, i. e.,

$$\tilde{\phi}(x) = \phi(x) \tag{2. 8. 2}$$

for almost all $x \in G$.

Proof. (i) According to (2. 7. 6), (2. 8. 1) is equivalent to

$$\Sigma\, d(U).\, \mathrm{Tr}[\hat{f}(U)\hat{\phi}(U)\hat{f}(U)^*] \geq 0 \tag{2. 8. 3}$$

for every $f \in C(G)$.

If each $\hat{\phi}(U)$ is p. s. a., one may (see Appendix A. 1. 4) write $\hat{\phi}(U) = T(U)T(U)^*$ for a suitable $T(U) \in \mathrm{End}\ (\mathscr{H}_U)$, in which case

$$\hat{f}(U)\hat{\phi}(U)\hat{f}(U)^* = (\hat{f}(U)T(U))(\hat{f}(U)T(U))^*$$

has a non-negative trace (see Appendix A. 2. 2) and (2. 8. 3) follows.

Reciprocally, if ϕ is PD, a suitable choice of $f \in C(G)$ in (2. 8. 3) leads to the conclusion that $\mathrm{Tr}[S^*S\hat{\phi}(U)] \geq 0$ for every $S \in \mathrm{End}\ (\mathscr{H}_U)$; in this connection, recall (2. 2. 5), (2. 6. 11) and (2. 6. 12). It then follows (Appendix A. 3. 1) that $\hat{\phi}(U)$ must be p. s. a.

(ii) This follows from (i), thanks to the uniqueness theorem (UT_1) in 2. 4. 1 combined with (2. 3. 3).

2.8.3. Continuous PD functions. It is not difficult to show (see Exercise 2.8.11) that a continuous complex-valued function ϕ on G is PD if and only if

$$\sum_{i,j=1}^{n} \phi(a_i^{-1}a_j)c_i\bar{c}_j \geq 0 \qquad (2.8.4)$$

for every finite sequence $(a_i)_{i=1}^n$ of elements of G and every finite sequence $(c_i)_{i=1}^n$ of complex numbers. (The passage between the double integral in (2.8.1) and the double sum in (2.8.4) rests upon the approximation of integrals by sums in standard fashion; cf. Exercise 2.1.18.)

The set of continuous positive definite functions on G will be denoted by P(G). It is evident that P(G) is a positive cone in C(G), i.e., the sum of two elements of P(G) lies in P(G), and the product by a non-negative real number of an element of P(G) lies in P(G).

The case $n = 2$ of (2.8.4) combines with a simple argument about Hermitian quadratic forms (see Exercise 2.8.11) to show that (2.8.2) holds for every $x \in G$, and that

$$|\phi(x)| \leq \phi(e) \qquad (2.8.5)$$

for every $x \in G$, whenever $\phi \in P(G)$. (As usual, e denotes the identity element of G.)

Remarks. (i) It follows from 2.8.2, (2.6.11) and (2.6.12) that for every $U \in \hat{G}$, the character χ_U is an element of P(G); as will be seen in 2.8.4(iii), they are the building bricks from which every central element of P(G) may be obtained in the form of a series; see also 2.11.1(b).

(ii) Except in the trivial case where G is finite, (2.8.5) does not hold, even merely for almost all $x \in G$, for every PD ϕ; in fact, if G is infinite, there exist many PD functions ϕ on G such that $\|\phi\|_\infty = \infty$. (For a construction of such functions, see Edwards [6].) □

For suitably restricted positive definite functions ϕ, including all those which are essentially bounded, it is possible to use formula (2.7.6) in such a way as to infer something about the pointwise convergence of the Fourier series of ϕ. As will appear in 2.9, this will form a useful step forward in handling the summability or convergence of

Fourier series of more general functions. The extra restriction on ϕ is the existence of a number $m = m_\phi$ and a heighbourhood $N = N_\phi$ of e such that

$$f * \phi * \tilde{f}(e) \le m \|f\|_1^2 \qquad (2.8.6)$$

for every $f \in C(G)$ with support

$$\operatorname{supp} f = \{x \in G : f(x) \neq 0\}^-$$

contained in N.

In 2.8.4 to follow, $\Sigma \ldots$ denotes $\Sigma_{U \in \hat{G}} \ldots$.

2.8.4. Theorem. (i) *Suppose that* ϕ *is PD and satisfies* (2.8.6). *Then*

$$\Sigma\, d(U).\operatorname{Tr} \hat{\phi}(U) < \infty \qquad (2.8.7)$$

and

$$\phi(x) = \Sigma\, d(U).\operatorname{Tr}[\hat{\phi}(U)U(x)^*] \qquad (2.8.8)$$

for almost all $x \in G$, *the series being absolutely and uniformly convergent for every* $x \in G$.

(ii) *If* $\phi \in P(G)$ *it satisfies* (2.8.6) *and* (2.8.7), *and* (2.8.8) *holds for every* $x \in G$.

(iii) *The central functions in* $P(G)$ *are precisely the functions of the form*

$$\phi = \Sigma\, c_U \chi_U , \qquad (2.8.9)$$

where the numbers c_U *are real and non-negative and such that*

$$\Sigma\, d(U) c_U < \infty , \qquad (2.8.10)$$

which condition ensures the absolute and uniform convergence of the series appearing in (2.8.9).

Proof. (i) Take (see 2.1.8) an approximate identity (k_j) in which every $k_j \in C(G)$ and vanishes on $G\backslash N$. By Exercise 2. 3. 5, $\lim_j \hat{k}_j(U) = I_U$ for every $U \in \hat{G}$. Also, by (2. 7. 6),

$$k_j * \phi * \tilde{k}_j(e) = \Sigma\, d(U).\, \mathrm{Tr}[\hat{k}_j(U)\hat{\phi}(U)\hat{k}_j(U)^*]\,,$$

every term of this series being non-negative by virtue of 2. 8. 2(i) and basic properties of Tr (see Appendix A. 2). On letting j increase and using (2. 8. 6), (2. 8. 7) follows without trouble. Then (see Appendix A. 1. 4, A. 1. 5 and (A. 2. 11)) the series on the right of (2. 8. 8) is seen to converge absolutely and uniformly; cf. the calculations in 2. 9. 2 and 2. 9. 4. Its sum-function ϕ_1 is therefore continuous and (by uniform convergence and the orthogonality relations; see especially (2. 6. 11) and (2. 6. 12)) $\hat{\phi}_1$ agrees with $\hat{\phi}$. Equality a. e. in (2. 8. 8) thus follows from the uniqueness theorem (UT_1) of 2. 4. 1.

(ii) It is evident that any $\phi \in P(G)$... indeed, any PD ϕ which is essentially bounded on some neighbourhood of e... satisfies (2. 8. 6). By (i), ϕ and the sum-function ϕ_1 agree a. e.; since both are continuous, they agree everywhere on G (see 2. 1. 4(i)).

(iii) Combine 2. 6. 7, 2. 8. 2 and (2. 2. 11) with (ii) immediately above.

Remarks. (i) As has been mentioned, (2. 8. 6) is satisfied by any ϕ which is essentially bounded on some neighbourhood of e. It is interesting to note that 2. 8. 4 implies that, conversely, if ϕ satisfies (2. 8. 6), then ϕ is essentially bounded.

(ii) Both 2. 8. 2 and 2. 8. 4 can be formulated so as to apply also to positive definite measures, the extension of 2. 8. 4 being interpreted by means of the alias described in 2. 1. 4(iii). However, little or no use will be made of these extensions.

(iii) It is quite simple to derive the Parseval formula (2. 7. 3) directly from 2. 8. 4; see Exercise 2. 8. 14. In this way, 2. 8. 4 could be made the basis of almost everything.

2. 8. 5. Elementary positive definite functions. Although it is evident that $P(G) + P(G) \subseteq P(G)$, it is equally evident that

$P(G) - P(G) \not\subseteq P(G)$. This feature, coupled with 2.8.2(i) and the usual partial ordering of s.a. endomorphisms of a Hilbert space (see Appendix A.0.2), suggests the consideration of a partial order $\ll$ on $P(G)$ defined as follows: $\psi \ll \phi$ if and only if $\phi, \psi \in P(G)$ and $\phi - \psi \in P(G)$. Once this is done, it seems natural to seek the elements ϕ of $P(G)$ which are minimal with respect to this partial order, i.e., which are non-zero and for which

$$\psi \in P(G), \quad \psi \ll \phi \Rightarrow \psi = c\phi$$

for some number $c \geq 0$. It is for some reason more usual to apply the term elementary positive definite function (briefly EPD function) to such minimal elements ϕ of $P(G)$.

In addition to this, an element ϕ of $P(G)$ is said to be normalised if and only if $\phi(e) = 1$. Notice that, by (2.8.5), every non-zero $\phi \in P(G)$ satisfies $\phi(e) > 0$, and $\phi(e)^{-1}\phi$ is normalised.

In the sequel, NEPD will often be written as an abbreviation for 'normalised elementary positive definite'.

It has been hinted in 2.0.2 that there are close and fundamental connections between unitary representations and PD functions. Although there is no room in these notes for a systematic development of this theme, at least some of the connections will now begin to emerge, and this in such a way as to add further point to the concept of EPD functions.

As a beginning, it is simple to check that, if U is any continuous unitary representation of G (not necessarily an element of $\hat{G}$), if $\underset{\sim}{e} \in \mathscr{H}_U$ satisfies $\|\underset{\sim}{e}\| = 1$, and if P is the orthogonal projection of $\mathscr{H}_U$ onto the subspace generated by $\underset{\sim}{e}$, then the function ϕ defined by the formula

$$\phi(x) = (\underset{\sim}{e} | U(x)\underset{\sim}{e}) = \mathrm{Tr}[PU(x)^*], \tag{2.8.11}$$

often termed a characteristic function of U, is a normalised element of $P(G)$. (The converse of this is discussed in 2.8.10(ii).) Also, unitarily equivalent representations give rise in this fashion to the same set of characteristic functions. As the next result shows, the NEPD functions are just those which arise as characteristic functions of continuous irreducible unitary representations of G.

2.8.6. Theorem. The NEPD functions on G are precisely those of the form (2.8.11), where $U \in \hat{G}$ and P is a one-dimensional (orthogonal) projector on $\mathcal{H}_U$.

Proof. Suppose ϕ is given by (2.8.11) for some $U \in \hat{G}$. Then $\hat{\phi}(U) = d(U)^{-1}P$ and $\hat{\phi}(V) = 0$ for $V \in \hat{G}$, $V \neq U$; see (2.6.11) and (2.6.12). So, if ψ and $\phi - \psi$ belong to P(G), 2.8.2 shows that $\hat{\psi}(V) = 0$ for $V \in \hat{G}$, $V \neq U$, and that $d(U)^{-1}P - \hat{\psi}(U)$ is positive self-adjoint. Since P is a one-dimensional projector, the last condition entails that $\hat{\psi}(U) = cP$ for some non-negative number c. Hence, by the uniqueness theorem of 2.4.1, ψ is a non-negative multiple of ϕ, showing that the latter is elementary.

On the other hand, if ϕ is a NEPD function, it is easy to infer from 2.8.2 and the uniqueness theorem that there is precisely one $U \in \hat{G}$ such that $\hat{\phi}(U) \neq 0$. Furthermore, the spectral theorem (see Appendix A.1) shows that $\hat{\phi}(U)$ must be a non-negative multiple of some one-dimensional projector. Normalisation and the uniqueness theorem then combine with (2.6.11) and (2.6.12) to establish (2.8.11).

Remarks. (i) It is a corollary of 2.8.6 that every elementary PD function ϕ belongs to T(G), and that T(G) could be (and sometimes is) defined to be the linear subspace of C(G) generated by all NEPD functions on G.

(ii) By using 2.8.6 and the argument appearing in the proof of 2.7.5(ii), it can be seen that the product of two EPD functions is a finite sum of such functions. This remark will be useful a little later. □

At this point it is possible to derive what is in fact the exact analogue of a theorem proved first by Bochner for certain PD functions on the group R.

2.8.7. Theorem (Bochner). The elements of P(G) are precisely the functions of the form

$$\phi = \sum_{i=1}^{\infty} c_i \phi_i , \qquad (2.8.12)$$

where the ϕ_i are NEPD functions on G and the c_i are non-negative numbers such that $\sum_{i=1}^{\infty} c_i < \infty$ (which ensures that the series in (2.8.12) converges absolutely and uniformly on G).

Proof. If the ϕ_i and c_i are as stated, it is evident that (2.8.12) defines an element ϕ of P(G). Conversely, suppose that $\phi \in P(G)$. In order to establish a representation of the type (2.8.12), it suffices in view of (2.8.8) to show that each function $x \mapsto \mathrm{Tr}[\hat{\phi}(U)U(x)^*]$ is expressible in the form (2.8.12). But, by 2.8.2(i), $\hat{\phi}(U)$ is positive self-adjoint, and the desired result follows at once from the spectral theorem (see Appendix A.1.2).

Remarks. (i) It is one of the less savoury aspects of non-Abelian existence that the expression (2.8.12) of a given $\phi \in P(G)$ is far from unique; see 2.11.1(c) below. Uniqueness obtains when G is Abelian; see 2.8.9 below.

(ii) In view of Remark (ii) following 2.8.6, 2.8.7 shows that

$$P(G).P(G) \subseteq P(G) . \qquad (2.8.13)$$

This, combined with 2.8.1 and Remark (i) following 2.8.6, could be used to recover 2.7.5(i) and (ii).

(iii) Note also that 2.8.4 alone, or 2.8.6 and 2.8.7 in combination, are enough to yield the completeness theorem (CTii) of 2.4.1.

2.8.8. **Return to the completeness theorem.** It is now possible to give the promised proof of the equivalence of (CTi) and (CTii) in 2.4.1.

Proof that (CTi) implies (CTii). It suffices to show that, if $a \in G$ and $a \neq e$ (the identity element of G), then $U(a) \neq I_U$ for some $U \in \hat{G}$. Now, one may easily construct a continuous PD function ϕ on G such that $\phi(e) = 1$ and $\phi(a) = 0$. For example, choose a symmetric neighbourhood N of e in G such that $a \notin N^2$, and define $\phi = g * \tilde{g}$, where g is a suitable scalar multiple of the characteristic function of N. It then follows from 2.8.4 that U(a) must differ from $U(e) = I_U$ for at least one $U \in \hat{G}$.

Proof that (CTii) implies (CTi). It is known from 2.7.5 that T(G) is a sub-algebra of C(G) which is stable under complex conjugation.

(CTii) asserts exactly that T(G) separates the points of G. So, by the Stone-Weierstrass theorem (see, for example, Edwards [2], Section 4.10), T(G) is dense in C(G). Since C(G) is dense in $L^2(G)$, (CTi) follows.

Remark. Given that T(G) is dense in C(G), it follows that T(G) is dense in $L^p(G)$ for any $p \in [1, \infty)$; this, too, might be referred to as the (or a) completeness theorem. More refined and constructive versions of this result stem from 2.9.2 and 2.9.3 below.

2.8.9. The Abelian case. In this case 2.8.2(i) says that an integrable function ϕ on G is positive definite if and only if the scalar-valued Fourier transform $\hat{\phi}$ of ϕ is non-negative, i.e., $\hat{\phi}(\chi) \geq 0$ for every $\chi \in \Gamma$; and 2.8.4 asserts that, if ϕ is positive definite and satisfies (2.8.6), then

$$\Sigma_{\chi \in \Gamma} \hat{\phi}(\chi) < \infty \qquad (2.8.14)$$

and

$$\phi(x) = \Sigma_{\chi \in \Gamma} \hat{\phi}(\chi)\chi(x) \qquad (2.8.15)$$

for almost every $x \in G$, with equality holding for every $x \in G$ whenever ϕ is continuous.

According to 2.8.6, the NEPD functions on G are none other than the continuous multiplicative characters of G (i.e., the elements of Γ), so that 2.8.7 in this case comes back to the validity of (2.8.15) for every $x \in G$ and every $\phi \in P(G)$.

2.8.10. General comments. (i) From 2.8.4 one may show (by arguments very similar to those used in 2.9.8 below) that the elements of P(G) are precisely the functions $f * \tilde{f}$ with $f \in L^2(G)$. (Incidentally, this may be combined with Remark (i) following 2.8.6 in such a way as to provide a proof of 2.9.8(iv).)

(ii) From 2.8.4 (or from 2.8.6 and 2.8.7 together) it is possible (see Exercise 2.8.13) to show that every $\phi \in P(G)$ is expressible in the form

$$\phi(x) = (\underset{\sim}{a} | U(x)\underset{\sim}{a}) , \tag{2.8.16}$$

where U is some continuous unitary representation of G and $a \in \mathscr{H}_U$; the representation U is in general infinite dimensional and reducible. In the Godement-Gelfand-Raikov theory mentioned in 2. 0. 2(iv), this connection between PD functions and unitary representations is established at the outset and is fundamental in all subsequent developments. The elementary PD functions then appear as those which correspond to irreducible representations; and the decomposition of ϕ described in 2. 8. 7 corresponds to the decomposition of a reducible representation into its irreducible components (much as in 2. 2. 5, except that the Godement-Gelfand-Raikov theory has to handle cases in which infinitely many components are present). In this theory, 2. 8. 7 is derived by a procedure quite different from that used in these notes... a procedure based on one of the big theorems of functional analysis (the Krein-Milman theorem; see, e. g., Edwards [2], Chapter 10).

2. 8. 11. Exercise. Write out detailed proofs of (2. 8. 4) and (2. 8. 5).

2. 8. 12. Let $\mathscr{H}$ be a f. d. Hilbert space with dimension d. Show that End ($\mathscr{H}$) is a Hilbert space relative to the scalar product $(A|B) = d. \mathrm{Tr}\, AB^*$.

2. 8. 13. Exercise. Prove that every $\phi \in P(G)$ is expressible in the form (2. 8. 16)

(a) by using 2. 8. 4;

and (b) by using 2. 8. 6 and 2. 8. 7 in combination.

(Exercises 2. 2. 16 and 2. 8. 12 should suggest a possible course of action.)

2. 8. 14. Exercise. Show how to derive the Parseval formula from 2. 8. 4.

2. 9. Summability and convergence of Fourier series

2. 9. 0. Preliminary remarks. The Fourier series to be considered from the point of view of convergence and summability are such as to make a few preliminary comments desirable.

Suppose given a set I and a complex-valued function $i \mapsto a_i$ with domain I. For any finite subset Δ of I, write $s_\Delta = \Sigma_{i \in \Delta} a_i$. If I is infinite, it is natural to suppose that the convergence of the series $\Sigma_{i \in I} a_i$, and its sum when it is convergent, are both to be defined in terms of the limiting behaviour of s_Δ as Δ expands so as to ultimately embrace every element of I. In other words, the said series will be said to be convergent and to have a sum s, if and only if to every $\varepsilon > 0$ corresponds a finite subset $\Delta_0 = \Delta_0(\varepsilon, (a_i))$ of I such that $|s_\Delta - s| \le \varepsilon$ for every finite subset Δ of I such that $\Delta \supseteq \Delta_0$. This definition is modified in the usual way, if $i \mapsto a_i$ is real-valued and s is $-\infty$ or ∞; it also generalises to the case in which $i \mapsto a_i$ takes its values in any Abelian topological group.

Although this definition is logically up to scratch, it proves to be too coarse to be of very much lasting value. This is because, in the complex-valued case, the series converges to a finite sum in the above sense, if and only if $\Sigma_{i \in I} |a_i| < \infty$, i. e., if and only if it converges absolutely (see, for example, Bourbaki, Topologie Générale, Chapter IV, §7, no. 2, p. 120). To get more refined concepts of convergence, one has to rely on some structural properties of I to select a suitable net (Δ_j) of finite subsets of I and consider the limiting behaviour of (s_{Δ_j}) as j increases. For example, if I is the set of positive integers, one takes the sequence (Δ_j) in which $\Delta_j = \{1, 2, \ldots, j\}$; and if I is the set $\mathbf{Z}$ of all integers, one usually takes the sequence (Δ_j) in which $\Delta_j = \{n \in \mathbf{Z} : |n| \le j\}$.

2.9.1. The case of Fourier series. Here one is confronted with the case in which $I = \hat{G}$, the corresponding finite partial sums being of the type

$$s_\Delta f = \Sigma_{U \in \Delta} d(U) \chi_U * f = D_\Delta * f , \qquad (2.9.1)$$

where Δ denotes a finite subset of $\hat{G}$, $f \in L^1(G)$ and

$$D_\Delta = \Sigma_{U \in \Delta} d(U) \chi_U \qquad (2.9.2)$$

is a sort of 'Dirichlet kernel' for G. [Actually, to take the Fourier series in the form (2.9.1) means that one is already grouping terms in a particular way: one has grouped together all the terms arising from one representation U; cf. 2.7.1.] The associated convergence problem (or problems) is therefore that (or those) concerned with the validity of the limiting relations

$$\lim s_\Delta f = f \tag{2.9.3}$$

in the sense of various topologies [for example, in the sense of any one of the normed spaces $C(G)$ and $L^p(G)$] and of the 'pointwise' relation

$$\lim s_\Delta f(x) = f(x) \tag{2.9.4}$$

for one or more elements x of G.

It has been seen in 2.7.2 and 2.7.3 that (2.9.3) holds in the sense of $L^2(G)$ for every $f \in L^2(G)$. As experience with the most familiar case $G = \mathbf{T}$ (see Edwards [3], Chapter 10) would lead one to expect, this is in general about the only simple, sweeping convergence theorem available. (Strange as it may at first seem, this familiar case is not the most favourable one; see 2.9.7 below.)

Within the context of Fourier theory, summability almost always amounts to replacing $s_\Delta f = D_\Delta * f$ by $k_j * f$, where (k_j) is an approximate identity characteristic of the particular summability method in question. Experience with the case $G = \mathbf{T}$ (see Edwards [3], Chapter 6) leads one to hope that summability will work more successfully than does convergence, a hope that shows every sign of being justified. For this reason, summability will be considered first. It provides one reasonably successful way of recapturing a function from its Fourier transform.

2.9.2. Take (see 2.1.8) an approximate identity (k_j) in which each k_j is non-negative, continuous, central, such that $\int k_j(x)dx = 1$, and such that k_j ultimately vanishes outside any preassigned neighbourhood of e. The same properties are shared by the functions $k_j * \tilde{k}_j$, which are continuous and positive definite. So it may and will be assumed that each k_j is continuous and positive definite. By 2.6.7 and 2.8.2, $\hat{k}_j(U) = c_j(U)I_U$, the number $c_j(U)$ being real and non-negative.

According to (2.8.8),

$$k_j(e) = \Sigma\, d(U).\, \mathrm{Tr}\, \hat{k}_j(U) = \Sigma\, d(U)^2 c_j(U) < \infty .$$

If $f \in L^1(G)$, the Fourier series of $k_j * f = f * k_j$ becomes

$$\Sigma\, d(U) c_j(U).\, \mathrm{Tr}[\hat{f}(U)U(x)^*] .$$

Herein we have (Appendix A, formulae (A. 2. 7) and (A. 2. 8)):

$$\begin{aligned} |\mathrm{Tr}[\hat{f}(U)U(x)^*]| &\le d(U)^{\frac{1}{2}} (\mathrm{Tr}\, \hat{f}(U)\hat{f}(U)^*)^{\frac{1}{2}} \\ &\le d(U) \|\hat{f}(U)\| \le d(U) \|f\|_1 , \qquad (2.9.5) \end{aligned}$$

so that the Fourier series in question converges absolutely and uniformly. Therefore

$$f * k_j(x) = k_j * f(x) = \Sigma\, d(U) c_j(U) \mathrm{Tr}[\hat{f}(U)U(x)^*] \qquad (2.9.6)$$

holds everywhere, the series converging absolutely and uniformly.

If we now let j vary, 2.1.9 affirms that the left-hand side of (2.9.6) converges to $f(x)$ at each point x of continuity of f. In particular, one has a process of summing uniformly the Fourier series of a continuous function; see also 2.9.6.

2.9.3. From 2.1.9 it follows that also $k_j * f$ converges to f in $L^p(G)$ for any $f \in L^p(G)$, if $1 \le p < \infty$, and weakly to f if $f \in L^\infty(G)$.

2.9.4. If k_j is replaced by any continuous positive definite function ϕ (not necessarily central), and if $f \in L^2(G)$, we can still show that the Fourier series of $f * \phi$ and of $\phi * f$ converge absolutely and uniformly. Take the latter, for example: we have (Appendix A, formulae (A. 2. 6) and (A. 2. 9))

$$\begin{aligned} &|\mathrm{Tr}\, \hat{\phi}(U)\hat{f}(U)U(x)^*| = |\mathrm{Tr}\, \hat{\phi}(U)^{\frac{1}{2}}.\, \hat{\phi}(U)^{\frac{1}{2}}\hat{f}(U)U(x)^*| \\ &\le (\mathrm{Tr}\, \hat{\phi}(U))^{\frac{1}{2}}.\, (\mathrm{Tr}\, \hat{\phi}(U)^{\frac{1}{2}}\hat{f}(U)U(x)^*U(x)\hat{f}(U)^*\hat{\phi}(U)^{\frac{1}{2}})^{\frac{1}{2}} \\ &= (\mathrm{Tr}\, \hat{\phi}(U))^{\frac{1}{2}}.\, \{\mathrm{Tr}\, \hat{\phi}(U)^{\frac{1}{2}}\hat{f}(U).\, (\hat{\phi}(U)^{\frac{1}{2}}\hat{f}(U))^*\}^{\frac{1}{2}} \\ &\le (\mathrm{Tr}\, \hat{\phi}(U))^{\frac{1}{2}}.\, (\mathrm{Tr}\, \hat{\phi}(U))^{\frac{1}{2}}.\, (\mathrm{Tr}\, \hat{f}(U)\hat{f}(U)^*)^{\frac{1}{2}} \\ &= \mathrm{Tr}\, \hat{\phi}(U).\, (\mathrm{Tr}\, \hat{f}(U)\hat{f}(U)^*)^{\frac{1}{2}} ; \end{aligned}$$

and, by the Parseval formula (2. 7. 3),

$$d(U)^{\frac{1}{2}} \{\mathrm{Tr}\, \hat{f}(U)\hat{f}(U)^* \}^{\frac{1}{2}} \leq \|f\|_2 .$$

Since $d(U) \geq 1$, the asserted absolute and uniform convergence follows from 2. 8. 4.

2. 9. 5. **Remark.** The argument given in 2. 9. 4 breaks down if one knows merely that $f \in L^1(G)$, in which case all that is evidently true is (see Appendix A, formula (A. 2. 8) that

$$\{\mathrm{Tr}\, \hat{f}(U)\hat{f}(U)^* \}^{\frac{1}{2}} \leq d(U)^{\frac{1}{2}} \|f\|_1 \quad ;$$

and there is presumably no reason to suppose that

$$\Sigma\, d(U)^{3/2} \mathrm{Tr}\, \hat{\phi}(U) < \infty .$$

More generally, it is natural to ask whether $f * g$ and $g * f$ have Fourier series which converge absolutely (or absolutely and uniformly) whenever the Fourier series of g has that property and $f \in L^1(G)$. When G is Abelian, the answer is (trivially) 'Yes'; otherwise the answer is unknown to the writer. Cf. 2. 9. 8 below.

2. 9. 6. The results of 2. 9. 2 and 2. 9. 3 show how to construct summability methods which are effective in the pointwise sense at points of continuity and in the L^p-sense for p finite.

In 2. 9. 2, instead of supposing that each k_j vanishes outside a small neighbourhood of e, one may assume that each is a t. p. (simply replace k_j by a suitable partial sum of its Fourier series, which is absolutely and uniformly convergent by 2. 8. 4). Then $k_j * f$ is a t. p. which, as j increases, converges uniformly to f if $f \in C(G)$ and converges in norm (or weakly, if $p = \infty$) in $L^p(G)$ to f if $f \in L^p(G)$. This includes a strong and fairly explicit version of the completeness theorem (CTi) in 2. 4. 1.

There remains the subtler question of the possibility of finding summability processes which are effective in the pointwise sense at almost all points of G for any given $f \in L^1(G)$. For a discussion of this, see Edwards and Hewitt [1]; see also Mayer [1].

2.9.7. Concerning convergence of Fourier series. As was remarked in 2.9.1, experience with the most familiar case $G = \mathbf{T}$, the circle group, leads one to expect that convergence presents more delicate problems than does summability. On top of this comes the fact that complications tend to gather thick and fast when one passes to general Abelian and then to non-Abelian groups. For one thing, neither Γ nor $\hat{G}$ can in general be ordered, and there is in general no obvious way in which terms of the Fourier series should be grouped so as to form a sequence or net of partial sums; cf. 2.9.0.

Notice that, even when $\hat{G}$ is countable (which is so if and only if G is first countable), being therefore the union of various sequences (Δ_j) of finite subsets Δ_j such that $\Delta_j \subseteq \Delta_{j+1}$, there is a priori no good reason for selecting any one such sequence rather than any other; nor is there any reason to suppose that the behaviour of the sequence $(s_{\Delta_j} f)$ for a selected sequence (Δ_j) is anything like that of the corresponding sequence of partial sums stemming from another choice of the sequence (Δ_j). One can hope that a judicious choice of (Δ_j) will lead to nice answers; but again the cast $G = T$ prevents one from expecting too much... except for sufficiently regular functions f.

As was seen in 2.7.2 and 2.7.3, (2.9.3) holds in the $L^2(G)$ sense whenever $f \in L^2(G)$. From 2.8.4 it follows that (2.9.3) holds in the sense of $C(G)$ (i.e., the sense of uniform convergence), whenever f is continuous and positive definite. It is natural to ask whether the formula (2.9.3) is valid for every $f \in L^p(G)$ for some $p \in [1, \infty]$, $p \neq 2$, or for $f \in C(G)$ and $p = \infty$. This may be considered in two stages according as G is or is not Abelian.

(i) If G is infinite and Abelian, the answer is 'No'. This may be established in various ways. For example, if the answer were 'Yes', it would (see 2.9.0) be the case that

$$\Sigma_{\chi \in \Gamma} |\hat{f}(\chi)\hat{g}(\chi)| < \infty \tag{2.9.7}$$

for every $f \in L^p(G)$ and every $g \in L^{p'}(G)$, where $1/p + 1/p' = 1$. This condition is symmetric in p and p' and so one may assume that $1 \leq p < 2$. Then (2.9.7) could be shown to entail that, for every $f \in L^p(G)$ and every function $\omega : \Gamma \to \mathbf{T}$, $\omega\hat{f}$ is the Fourier transform of some element of $L^p(G)$. However (see Edwards [3], 14.3.5... a result which can be extended without trouble to any infinite compact Abelian G), it would then ensue that $L^p(G) \subseteq L^2(G)$, which is false since G is infinite compact (see Exercise 2.1.16 above).

(ii) If G is non-Abelian, the answer is still generally 'No', though the writer is not sure whether this is the case for every infinite compact G.

In spite of this rebuff, it can happen that, for suitably chosen sequences or nets (Δ_j) of finite subsets of G which expand to cover G, one has

$$\lim s_{\Delta_j} f = f$$

in $L^p(G)$ for every $f \in L^p(G)$ when $1 < p < \infty$. For the case $G = \mathbf{T}$, see Edwards [3], 12.10.1. Nothing like this can happen for $G = \mathbf{T}$ when $p = 1$ or when $p = \infty$ (even if, in the latter case, one handles only continuous functions f); in this connection the situation which is nicest (apart from the trivial case when G is finite) is that in which G is zero-dimensional (as described in 2.2.13, for example); for details see Edwards and Price [1], §9.

Lack of space forbids further pursuit of what by now is fairly evidently a complicated business.

2.9.8. **Absolute convergence of Fourier series; the algebra A(G).** It has emerged in 2.8.4 that the Fourier series of f is absolutely and uniformly convergent whenever f is continuous and positive definite. Beside this, it is obvious that, if G is Abelian, the Fourier series of f is absolutely convergent if and only if

$$\Sigma_{\chi \in \Gamma} |\hat{f}(\chi)| < \infty, \tag{2.9.8}$$

in which case the series is obviously uniformly convergent as well (but see Exercises 2.9.13 and 2.14.9).

In the non-Abelian case, it is not clear at the outset that there is any condition which (like (2.9.8)) is simply expressible in terms of $\hat{f}$ and which is necessary and sufficient to ensure the absolute (or absolute and uniform) convergence of the Fourier series of f. There are however, two useful pointers in the shape of (a) and (b) immediately below.

(a) One may revert to the Abelian case and note that (2. 9. 8) is equivalent to the condition that $\hat{f}$ be the pointwise product of two functions in $l^2(\Gamma)$; and that (by the Parseval formulae) this is equivalent to the condition that f be equal almost everywhere to a function of the type $g * \tilde{h}$ (or $g * h$), where g and h belong to $L^2(G)$ (cf. Edwards [3], p. 168, Remark (4)).

(b) One may try to mimic the condition (2. 9. 8) itself. To do this it is necessary to bear in mind that, if $\mathscr{H}$ denotes any f. d. Hilbert space and $A \in \mathrm{End}(\mathscr{H})$, then there is (see Appendix A. 1. 5) a unique p. s. a. endomorphism $|A| \in \mathrm{End}(\mathscr{H})$ such that $|A|^2 = AA^*$, and that (as one would expect) $A \mapsto |A|$ has at least some of the properties of the absolute value function defined for complex numbers. It is, for example, true that $A \mapsto \mathrm{Tr}\,|A|$ is a norm on $\mathrm{End}(\mathscr{H})$: this is not trivial, but it is a consequence of the important formula

$$\mathrm{Tr}\,|A| = \sup\{|\mathrm{Tr}\,AX| : X \in \mathrm{End}(\mathscr{H}),\ \|X\| \leq 1\}, \qquad (2.9.9)$$

valid for every $A \in \mathrm{End}(\mathscr{H})$; see Appendix A, formula (A. 2. 10).

It is not very surprising that a plausible analogue of (2. 9. 8) is expressed by the condition

$$\|f\|_A < \infty,$$

where

$$\|f\|_A = \Sigma_{U \in \hat{G}}\, d(U).\,\mathrm{Tr}\,|\hat{f}(U)| \ (\leq \infty) \qquad (2.9.10)$$

for every $f \in L^1(G)$. The function $f \mapsto \|f\|_A$ satisfies the conditions

$$\|f_1 + f_2\|_A \leq \|f_1\|_A + \|f_2\|_A,$$
$$\|\alpha f\|_A = |\alpha|\,\|f\|_A,$$

for f_1, f_2, $f \in L^1(G)$ and α a complex number, provided one adopts the conventions described in 1. 3. 3 above. It follows that $\{f \in L^1(G) : \|f\|_A < \infty\}$ is a linear subspace of $L^1(G)$, the restriction of $\|.\|_A$ to which is a norm

(One might have defined $|A|$ to be the positive square root of A^*A, rather than that of AA^*; however the two positive square roots can be shown to have the same trace, so that $\|.\|_A$ remains unaffected; see Appendix A, formula (A. 2. 12).

With these comments in mind, one may now (whether or not G is Abelian) consider the following statements:

(i) $f = g * h$ a. e. for suitably chosen $g, h \in L^2(G)$;

(ii) $f \in L^1(G)$ and $\|f\|_A < \infty$;

(iii) the Fourier series of f converges absolutely and uniformly.

On the basis of what has already been covered in these notes it is possible to show that (i) and (ii) are equivalent; see Exercise 2. 9. 9. Moreover, reference to (2. 9. 9) makes it easy to verify that (ii) implies (iii); and it is not difficult to show (without reference to (ii)) that (i) implies (iii); see Exercise 2. 9. 10.

However, although in the Abelian case (i), (ii) and (iii) are equivalent, there is an example (Mayer [1], Theorem 4. 1) to show that in general (iii) does not imply the other two. (The notation used in Mayer's paper, although apparently like that used in these notes, differs materially from the latter.)

At all events, any f satisfying (i) or (ii) (and hence (iii)) is equal a. e. to a continuous function. The set of $f \in C(G)$ satisfying (i) or (ii) is usually denoted by A(G). In the Abelian case, A(G) is precisely the set of continuous functions on G having absolutely convergent Fourier series; in general, A(G) is a proper subset of the set of continuous functions having absolutely and uniformly convergent Fourier series.

It is not difficult to verify that A(G) is the linear subspace of C(G) generated by P(G). As a result, A(G) is in fact a subalgebra of C(G) closed under complex conjugation; cf. 2. 8. 1 and (2. 8. 13). Also, A(G) is a Banach space when equipped with the norm $\|.\|_A$; see Exercise 2. 9. 11.

The inequality

$$\|fg\|_A \leq \|f\|_A \|g\|_A , \qquad (2. 9. 11)$$

which is very simple to prove when G is Abelian, remains valid in the non-Abelian case as well. In this case, perhaps the simplest proof is based upon the following characterisation of A(G) [which is closely analogous to that of P(G) mentioned in 2.8.10(ii)]:

(iv) A(G) consists precisely of those functions f of the form

$$f(x) = (\underset{\sim}{a} | U(x)\underset{\sim}{b}) , \tag{2.9.12}$$

where U denotes a continuous enitary representation of G and $\underset{\sim}{a}, \underset{\sim}{b} \in \mathscr{H}_U$; moreover

$$\|f\|_A = \min \|\underset{\sim}{a}\| . \|\underset{\sim}{b}\| , \tag{2.9.13}$$

the minimum being taken with respect to all possible representations (2.9.12) of f.

[The representations referred to in (iv) may be infinite-dimensional; they are not necessarily elements of, or equivalent to elements of, $\hat{G}$; see Exercise 2.9.12.] Granted (iv), the proof of (2.9.11) is quite simple [one uses tensor products, as in the proof of 2.7.5(ii)].

A(G) is thus a commutative Banach algebra with an identity (unit) element $\underline{1}$, the study of which is an important component of harmonic analysis. For further remarks on the Abelian case, see Edwards [3], Chapter 10; for the non-Abelian (compact) case, see Hewitt and Ross [1], §34. A lengthy and detailed study of A(G) applying to the case of general locally compact groups is to be found in Eymard [1].

2.9.9. **Exercise.** Prove that 2.9.8(i) is equivalent to 2.9.8(ii).

2.9.10. **Exercise.** Verify that 2.9.8(ii) implies 2.9.8(iii). Also, without reference to 2.9.8(ii), prove that 2.9.8(i) implies 2.9.8(iii).

2.9.11. **Exercise.** Write out a detailed proof of the fact that A(G) is a Banach space relative to the norm $\|.\|_A$.

2.9.12. **Exercise.** Prove 2.9.8(iv).

2.9.13. **Exercise.** Find an example (or examples) of functions $f \in C(\mathbf{T})$ whose Fourier series converge uniformly but not absolutely.

Remark. This shows in particular that $A(\mathbf{T}) \neq C(\mathbf{T})$. In fact $A(G) \neq C(G)$ for every infinite compact G; for the Abelian case, see Exercise 2.14.9 below; otherwise, see Hewitt and Ross [1], (37.4).

2.9.14. Exercise. Let G be a compact Abelian group with dual group Γ (see 2.5.4). Define Γ_∞ to be the set

$$\{\underline{1}\} \cup \{\chi \in \Gamma : \chi \text{ is of infinite order}\},$$

where $\underline{1}$ stands for $\underline{1}_G$, the identity element of Γ. Let f and g be trigonometric polynomials on G and suppose that

$$\{\chi \in \Gamma : \hat{g}(\chi) \neq 0\} \subseteq \Gamma_\infty .$$

Prove that

$$\lim_{n \in \mathbf{Z},\ |n| \to \infty} \int f(x) g(x^n) dx = \hat{f}(\underline{1})\hat{g}(\underline{1}) . \qquad (2.9.14)$$

Discuss the validity of (2.9.14) for more general functions f and g.

Remarks. For $G = \mathbf{T}$, one extended version of (2.9.14) is usually known as Fejér's lemma; see Edwards [3], Exercise 2.16. It is known (see Hewitt and Ross [1], (24.25)) that $\Gamma_\infty = \Gamma$ if and only if G is connected.

2.10. Closed spans of translates

2.10.0. It was asserted in 2.1.10 that harmonic analysis on G would be found to be intimately connected with the study of the structure of closed (translation-)invariant subspaces of such standard function spaces as C(G) and $L^p(G)$. In this and the next two sections, some detailed support for this statement will be exhibited.

The final aim is the complete characterisation of closed invariant subspaces and closed ideals in the above function spaces in terms of the Fourier transforms of members of the subspaces. In this section and the next, attention will be focused on closed invariant subspaces with a single generator: more could be achieved with little further effort but, for reasons mentioned in 2.10.7, it seems more convenient and economical in the long run to proceed somewhat differently and to defer until 2.12 a common treatment of closed invariant subspaces and closed ideals.

2.10.1. Let E be any one of the usual spaces C(G) and $L^p(G)$ $(1 \leq p < \infty)$ with the customary normed topology, or $L^\infty(G)$ with its weak topology as the dual of $L^1(G)$. If $f \in E$, we denote by LE(f) (resp. RE(f)) the closed linear subspace of E generated by the left (resp. right) translates $L_a f$ (resp. $R_a f$) as a varies over G; these subspaces are often termed the closed spans (in E) of the left (or right) translates of f. The immediate objective is to examine these subspaces by using the Hahn-Banach theorem. In what follows, E' denotes the topological dual of E and $\langle , \rangle$ the pairing between E and E'.

2.10.2. It is known that any continuous linear functional on E can be written as

$$g \mapsto \langle g, \nu \rangle = \int g(x^{-1}) d\nu(x) , \qquad (2.10.1)$$

where: $\nu \in M(G)$ if $E = C(G)$, and $\nu = \mu^h$ (see 2.1.4(iii)) for some $h \in L^{p'}(G)$ if $E = L^p(G)$, p' being defined as usual by $1/p + 1/p' = 1$; when $E = C(G)$ this is the substance of 1.2.2; when $E = L^p(G)$, see Edwards [3], Appendix C. Accordingly the topological dual E' is identified with M(G) or $L^{p'}(G)$ in such a way that

$$\langle L_{a^{-1}} f, \nu \rangle = f * \nu(a) , \qquad (2.10.2)$$

$$\langle R_{a^{-1}} f, \nu \rangle = \nu * f(a) , \qquad (2.10.3)$$

each of $f * \nu$ and $\nu * f$ being a continuous function on G whenever $f \in E$ and $\nu \in E'$; see 2.1.5. These equations combine with the completeness theorem, (2.3.4) and (2.3.5) to inform us that if $f \in E$ and $\nu \in E'$, then ν annihilates LE(f) (resp. RE(f)) if and only if $\hat{f}(U)\hat{\nu}(U) = 0$ (resp. $\hat{\nu}(U)\hat{f}(U) = 0$) for every $U \in \hat{G}$.

2.10.3. On the other hand, it is clear that if $f, g \in E$, then $g \in LE(f)$ (resp. RE(f)) if and only if $LE(g) \subseteq LE(f)$ (resp. $RE(g) \subseteq RE(f)$)

Putting all these facts together, the Hahn-Banach theorem (Edwards [3], Appendix B.5) leads via the substance of 2.3 and 2.4 to

2.10.4. Theorem. The notation being as above, let $f \in E$. In order that $g \in E$ shall belong to $LE(f)$ (resp. $RE(f)$), it is necessary and sufficient that the following condition be fulfilled:

$$\left.\begin{array}{l}\text{If } \nu \in E' \text{ and } \hat{f}(U)\hat{\nu}(U) = 0 \text{ (resp. } \hat{\nu}(U)\hat{f}(U) = 0) \\ \text{for all } U \text{ in } \hat{G}, \text{ then } \hat{g}(U)\hat{\nu}(U) = 0 \text{ (resp.} \\ \hat{\nu}(U)\hat{g}(U) = 0) \text{ for all } U \text{ in } \hat{G}.\end{array}\right\} \qquad (2.10.4)$$

It is useful to supplement 2.10.4 by other necessary and sufficient conditions.

2.10.5. Theorem. In order that $g \in E$ shall belong to $LE(f)$ (resp. $RE(f)$), f being a given element of E, it is necessary and sufficient that, for each $U \in \hat{G}$, $\hat{g}(U)$ shall be a left (resp. right) multiple of $\hat{f}(U)$:

$$\hat{g}(U) = M\hat{f}(U) \quad (\text{resp. } \hat{g}(U) = \hat{f}(U)M) , \qquad (2.10.5)$$

where M is an endomorphism of $\mathscr{H}_U$.

Proof. The sufficiency follows from 2.10.4 itself. As for necessity, consider the 'left' case as typical. For any h which is a finite linear combination of left-translates of f, say

$$h = \Sigma_p \lambda_p L_{a_p} f ,$$

we have

$$\hat{h} = \Sigma_p \lambda_p U(a_p)\hat{f}(U) = (\Sigma_p \lambda_p U(a_p))\hat{f}(U) ,$$

which is a left multiple of $\hat{f}(U)$. On the other hand, if h tends to g in E, then $\hat{h}(U) \to \hat{g}(U)$ for each U. Whence the necessity of the condition, in view of Lemma A.3.4 of Appendix A.

2.10.6. The Abelian case. Here there is no distinction between $LE(f)$ and $RE(f)$ and they will be denoted indifferently by $TE(f)$ (the 'T' to remind one of 'translates').

From 2.10.5 it follows that, if $f \in E$, a given $g \in E$ belongs to $TE(f)$ if and only if

$$\chi \in \Gamma, \quad \hat{f}(\chi) = 0 \Rightarrow \hat{g}(\chi) = 0 \tag{2.10.6}$$

It is also just as simple to prove (see Edwards [3], Section 11. 2) something more general, namely: suppose S is any non-void subset E and denote by TE(S) the closed invariant subspace of E generated by S; then TE(S) comprises exactly those $g \in E$ such that

$$\chi \in \Gamma, \quad \hat{f}(\chi) = 0 \quad (\forall\ f \in S) \Rightarrow \hat{g}(\chi) = 0 . \tag{2.10.7}$$

2. 10. 7. As was stated in 2. 10. 0, it would be feasible at this point to use 2. 10. 4 and 2. 10. 5 as the basis for discussing the structure of general closed invariant subspaces of E; cf. the assertion of which (2. 10. 7) forms part. However, it turns out that these closed invariant subspaces are the same as the closed ideals relative to convolution (see 2. 12. 1 below). Since the 'ideal' point of view is basic in several approaches, it seems more economical to cover both topics under the latter heading in 2. 12.

Before coming to the details of this programme, it may be worthwhile to consider some such structural questions in a more concrete guise. This will be done in 2. 11.

2. 10. 8. Exercise. Let G be compact Abelian, and let E be as in 2. 10. 1. Given $f \in E$ and $A \subseteq G$, denote by TE(f, A) the closed linear subspace of E generated by $\{L_a f : a \in A\}$. The translates of f are said to be independent in E (see Edwards [7]) if and only if $f \notin$ TE(f, A) for every closed subset A of G not containing e (the identity element of G).

Assume that G is first countable (i. e., that there exists a countable base of neighbourhoods of e in G). Show how to construct elements f of E such that

(i) TE(f) = TE(f, G) = E

and

(ii) the translates of f are independent in E.

Note. The above concept of independence of translates of f in E implies, but is generally stronger than, that of linear independence in the algebraic sense (this last is equivalent to saying that $f \notin TE(f, A)$ for every finite subset A of G not containing e). As defined above, independence is in fact a sort of topological linear independence.

2.10.9. **Exercise.** Let G be a connected compact Abelian group. Let $f \in \mathscr{A}(G)$, where $\mathscr{A}(G)$ is as defined in Exercise 2.7.10. Prove that TE(f, U) = TE(f) for every non-void open subset W of G.

Remark. The translates of f are about as far removed as possible from being independent in the sense of the preceding exercise: they are, one might say, 'madly intertwined'.
[Hints: Use Exercise 2.7.10, coupled with an appeal to the Hahn-Banach theorem (Edwards [3], Appendix B.5) and the substance of 2.10.2.]

2.11. Structural building bricks and spectra

2.11.1. Before embarking on the description of closed invariant subspaces and closed ideals, it may be worthwhile to glance rapidly at structural problems from a more concrete point of view.

Classical analysis may suggest consideration of the following sort of question. Is there any set Ω of reasonably simple and well-behaved functions ω on G, the elements of which may be termed 'base functions' and which act as natural 'building bricks' for the synthesis of all sufficiently well-behaved functions f on G? One might hope, too, that the elements of Ω are to be in some way distinguished from the point of view of harmonic analysis; and that likewise those base functions which figure in the synthesis of a given f shall be in some way related to the 'harmonic properties' of f.

The preceding description is very vague, and it will be left so. We add merely that the type of synthesis referred to is to be taken to be accomplished by the systematic formation of linear combinations of base functions, followed by taking limits of these combinations. (Illustrations (a) and (b) below exemplify what we have in mind.)

As a beginning, we review some relevant facts already established.

(a) If G is Abelian, everything works out nicely. It suffices to take $\Omega = \Gamma$ (the set of continuous multiplicative characters of G). The substance of 2.9 confirms that every well-behaved f is represented by its Fourier series

$$\Sigma_{\chi \in \Gamma} \hat{f}(\chi)\chi ,$$

and the orthogonality relations ensure that there are various senses in which this Fourier series of f is <u>the</u> proper series of multiples of characters to be used (see the discussion in Edwards [3], Chapter 1). Moreover, it appears from 2.10.6 that those χ which actually appear in the Fourier series of f (i.e., those for which $\hat{f}(\chi)$ is non-zero) are precisely those which are limits of linear combinations of translates of f.

(b) If one drops the assumption that G be Abelian, the substance of (a) remains valid on taking Ω to comprise all the characters χ_U $(U \in \hat{G})$, <u>provided</u> only central functions f are contemplated; see 2.6.7, 2.8.4(iii) and the substance of 2.9 and 2.10. However, it is plain that only central functions f can result from the synthesis (by limits of linear combinations) of characters χ_U, so that this choice of Ω is not a great success.

(c) If we persist in taking G non-Abelian and hope to handle non-central functions f, a different choice of Ω is essential. In view of (a) and (b), together with the fact that linear combinations of NEPD functions suffice to yield all trigonometric polynomials (see Remark (i) following Theorem 2.8.6), a reasonable choice for Ω would seem to be the set Φ of all NEPD functions on G. This choice is also suggested by Theorem 2.8.7, and yet again by recalling from 2.8.9 that $\Phi = \Gamma$ in case G is Abelian.

Now the results of 2.9 do indeed show that to every well-behaved function f on G corresponds at least one complex-valued function c on Φ such that

$$f = \Sigma_{\phi \in \Phi} c(\phi)\phi , \tag{2.11.1}$$

which is a promising start. But now difficulties arise in at least two forms, namely:

(i) there is in general no uniqueness in (2.11.1), i.e., the function c is in general not uniquely determined by f (cf. Remark (i) following 2.8.7);

(ii) it is generally impossible to find a representation (2.11.1) in which c has the property that $c(\phi) \neq 0$ implies $\phi \in LE(f)$ [or $RE(f)$]; in fact, the set of $\phi \in \Phi$ such that $c(\phi) \neq 0$ does not seem to be related in any intrinsic fashion to f.

These points are illustrated in Exercises 2.11.6 and 2.11.7. Of these two shortcomings, (i) is perhaps the more serious. At all events, the elements of Φ have only limited success in the role of base functions. Nor does there seem to be any choice which works well in all directions.

In spite of this setback, it will appear in 2.12 that the elements of Φ do suffice as building bricks, though in a way less direct and explicit than has been contemplated in the above discussion. Thus, 2.12.4 will affirm that every f is the limit of linear combinations of left (resp. right) translates of those ϕ which belong to $\Phi \cap LE(f)$ [resp. $\Phi \cap RE(f)$].

Reverting to (ii) above, it is possible to prove a partial substitute for restricted functions f, and this partial result will prove useful in 2.12.

2.11.2. Let E be as in 2.10.1 and let $f \in E$ be normal, i.e.,

$$f * \tilde{f} = \tilde{f} * f \tag{2.11.2}$$

Then f is the limit in E of trigonometric polynomials

$$\theta = \Sigma c_i \phi_i \tag{2.11.3}$$

in which $\phi_i \in \Phi \cap LE(f) \cap RE(f)$; moreover the ϕ_i may be chosen to be two-by-two orthogonal:

$$\phi_i * \phi_j = \phi_i * \tilde{\phi}_j = 0 \quad \text{whenever } i \neq j \tag{2.11.4}$$

and to be such that

$$\phi_i * f = f * \phi_i = c_i' \phi_i \tag{2.11.5}$$

for certain non-zero numbers c_i'. (Any normal trigonometric polynomial is in fact identical with some θ of the type just specified.)

Proof. In view of the results of 2.9, it is enough to consider the case in which f has the form $x \mapsto \mathrm{Tr}\,[AU(x)^*]$, where $U \in \hat{G}$ and $A \in \mathrm{End}\,(\mathscr{H}_U)$ is normal, i.e., $AA^* = A^*A$. In this case the desired result follows from Lemma A.3.2 of Appendix A in conjunction with Theorems 2.8.6 and 2.10.4.

2.11.3. Remarks. Every element of P(G) is normal. Also, every function of the form $f * \tilde{g}$ with $f, g \in L^2(G)$ (in particular, every element of A(G)) is a linear combination of normal functions of the type $h * \tilde{h}$ (take h in turn to be $f \pm g$ and $f \pm ig$).

From 2.11.2 it follows that if $f \in E$ is normal and different from $\underline{0}$, each of $\Phi \cap LE(f)$ and $\Phi \cap RE(f)$ is non-void. In addition, if $f \neq \underline{0}$ (normal or not), then $g = \tilde{f} * f$ is normal and different from $\underline{0}$; moreover, by the statement following (2.1.7), $g \in LE(f)$, hence $LE(g) \subseteq LE(f)$, and so $\Phi \cap LE(f)$ is non-void. Similarly, by considering the function $g' = f * \tilde{f}$, one sees that $\Phi \cap RE(f)$ is non-void. So we derive

2.11.4. Theorem (Beurling). _If_ $f \in E$ _and_ $f \neq \underline{0}$, _each of_ $\Phi \cap LE(f)$ _and_ $\Phi \cap RE(f)$ _is non-void._

Beurling's original theorem applied to the case in which G is the additive group **R** of real numbers and $E = L^\infty(G)$ with its weak topology as the dual of $L^1(G)$, though he also considered the case in which E is a space of uniformly continuous functions with a certain so-called 'narrow topology'. The L^∞-case was afterwards extended to arbitrary locally compact Abelian G.

All the results of §18 of Godement [1] for the compact case are implied by Theorems 2.11.2 and 2.11.4.

Some writers (Godement [1] for example) speak of $\Phi \cap LE(f)$ and $\Phi \cap RE(f)$ as the _left E-spectrum_ and the _right E-spectrum_ of f, respectively. Thus 2.11.4 asserts that every non-zero $f \in E$ has non-void spectra. It is quite easy to show that the E-spectra of $\phi \in \Phi$ are each identical with $\{\phi\}$. However, except when G is Abelian, the E-spectra of a normal trigonometric polynomial (2.11.3) do not coincide with the set of ϕ_i for which $c_i \neq 0$.

2.11.5. The Abelian case. Here there is no distinction between left and right spectra of f, which will be referred to simply as the E-spectrum of f.

In this case, 2.11.4 says that, for any non-zero $f \in E$, the E-spectrum of f contains at least one continuous multiplicative character $\chi \in \Gamma$. If $f \in E$, its E-spectrum is (by the substance of 2.10.6) the set

$$\{\chi \in \Gamma : \hat{f}(\chi) \neq 0\} ,$$

which is independent of E (assumed to be as in 2.10.1).

2.11.6. Exercise. Illustrate 2.11.1(i) by taking $f = \chi_U$, where $U \in \hat{G}$ and $d(U) > 1$.

2.11.7. Exercise. Let f be a trigonometric polynomial such that (2.11.1) holds for at least one $c : \Phi \mapsto \mathbf{C}$ having a finite support and such that $c(\phi) \neq 0$ implies $\phi \in LE(f)$. Show that

$$f \in \tilde{f} * T(G) \text{ and } \tilde{f} \in T(G) * f . \tag{2.11.6}$$

Assuming that G is non-Abelian, give an example of a trigonometric polynomial f on G which does not satisfy (2.11.6).

2.12. Closed ideals and closed invariant subspaces

2.12.0. Once again, E is to be as described in 2.10.1. However, more emphasis will now be placed on the structure of E as an algebra with respect to convolution and on the expression of the basic features of harmonic analysis in terms of the ideal structure of the algebra E: this is the viewpoint adopted in Loomis [1], Sections 27, 39, 40; see also Hewitt and Ross [1], §38.

In view of the remarks in 2.10.0 and 2.10.7, it is important that at the same time the relationship between closed ideals and closed invariant subspaces be cleared up. This will be done at the outset.

2.12.1. Preliminaries concerning ideals. By a <u>left</u> (resp. <u>right</u>) <u>ideal</u> in E is meant a linear subspace I of E with the property that

$$E * I \subseteq I \quad (\text{resp. } I * E \subseteq I),$$

where, if A, B, ..., C are subsets of E, $A * B * \ldots * C$ denotes the set of functions $a * b * \ldots * c$ obtained as a, b, ..., c range independently over A, B, ..., C, respectively. A <u>(two-sided) ideal</u> in E is a linear subspace I which is both a left ideal and a right ideal, i.e., which is such that

$$(E * I) \cup (I * E) \subseteq I.$$

An ideal of a given type (left, right or two-sided) need not be an invariant subspace of the same type; and vice versa. However, it is not difficult to show that a closed subset I of E is an ideal of given type, if and only if it is an invariant subspace of the same type. (The proof can be effected much as in Edwards [3], Section 11.1; the details will be left as Exercise 2.12.16 for the reader.)

In particular, then, LE(f) (resp. RE(f)) is the smallest closed left (resp. right) ideal in E containing f; and this proves to be none other than the closure in E of the left (resp. right) ideal $E * f$ (resp. $f * E$); see Exercise 2.12.17.

Part of the aim is to decompose E and its ideal into direct sums of minimal ideals, a left (resp. right, two-sided) ideal I being termed <u>minimal</u> if and only if $I \neq \{\underline{0}\}$ and I properly contains no left (resp. right, two-sided) ideal other than $\{\underline{0}\}$. The first step in this direction is to track down the minimal ideals in E, in doing which the following remarks may be found to be suggestive.

Consider left ideals for definiteness. If I is any such ideal, it contains $E * f$ for every $f \in I$. For any $f \in E$, $E * f$ is a left ideal in E; it contains f if and only if f is a t.p.; but in any case the closure of $E * f$ contains f and (see Exercise 2.12.17) is indeed the smallest closed left ideal containing f. [As is shown in the same exercise, the smallest left ideal containing f is the set of elements $g * f + \lambda f$ obtained when g ranges over E and λ over complex numbers; this is usually spoken of as the <u>left ideal generated by</u> f.] In view of this, it seems reasonable to conjecture that any minimal left ideal will be of the form $E * f$, where f is a suitable t.p. $\neq \underline{0}$. As will appear in 2.12.3, this is indeed the case.

It will turn out to be convenient to denote by $\phi_{U,P}$ the NEPD function $x \mapsto \mathrm{Tr}\,[PU(x)^*]$ (cf. 2.8.5 above) and to write $L_{U,P} = E * \phi_{U,P}$ and $R_{U,P} = \phi_{U,P} * E$. (Herein U is understood to denote an element of $\hat{G}$ and P a one-dimensional orthogonal projector on $\mathscr{H}_U$.) Each of $L_{U,P}$ and $R_{U,P}$ is finite dimensional and therefore closed, so that $L_{U,P}$ (resp. $R_{U,P}$) is the smallest left (resp. right) ideal containing $\phi_{U,P}$. Since $\phi_{U,P}$ is idempotent (i.e., $\phi_{U,P} * \phi_{U,P} = \phi_{U,P}$), $L_{U,P}$ is the set of $f \in E$ such that $f * \phi_{U,P} = f$; and similarly for $R_{U,P}$, with $\phi_{U,P} * f$ in place of $f * \phi_{U,P}$.

2.12.2. Lemma. Let $I \neq \{\underline{0}\}$ be a left (resp. right) ideal in E. Then $I \supseteq L_{U,P}$ (resp. $R_{U,P}$) for some $U \in G$ and some one-dimensional orthogonal projector P on $\mathscr{H}_U$.

Proof. Take the 'left' case. Choose $f \neq \underline{0}$ in I and consider $g = \tilde{f} * f$ (in the right-handed case, take $g = f * \tilde{f}$). Then $g \in I$ and $g \neq \underline{0}$; and, by Exercise 2.1.15, $g = \tilde{g}$. From 2.11.2 it follows that U and P exist such that $\phi_{U,P} = \text{const.}\ g * \phi_{U,P} = \text{const.}\ \phi_{U,P} * g$, whence it appears that $\phi_{U,P} \in I$. But then I must contain $L_{U,P}$ (the smallest left ideal containing $\phi_{U,P}$). The proof in the 'right' case is exactly analogous.

2.12.3. Theorem. A left (resp. right) ideal I is minimal if and only if it is of the form $L_{U,P}$ (resp. $R_{U,P}$).

Proof. By 2.12.2, I contains some $\phi_{U,P}$. But then I contains $L_{U,P}$ (resp. $R_{U,P}$). Since this last is a left (resp. right) ideal $\neq \{\underline{0}\}$, minimality shows that $I = L_{U,P}$ (resp. $I = R_{U,P}$).

Conversely, let $I = L_{U,P}$ (resp. $I = R_{U,P}$). It is clear that I is a left (resp. right) ideal $\neq \{\underline{0}\}$. Let I' be any left (resp. right) ideal contained properly in I: it has to be shown that $I' = \{\underline{0}\}$. Now if $f \in I'$, then $\hat{f}(V) = 0$ for $V \neq U$ and $\hat{f}(U) = AP$ (resp. $\hat{f}(U) = PA$) for some endomorphism A of $\mathscr{H}_U$. If I' were $\neq \{\underline{0}\}$, one could choose $f \neq \underline{0}$, so that $AP \neq 0$ (resp. $PA \neq 0$). But then Lemma A.3.6 of Appendix A would show that $I \subseteq I'$, contrary to hypothesis.

2.12.4. Theorem. Let I be a left (resp. right) ideal in E. Then I is contained in the closure in E of the direct sum of mutually orthogonal minimal left (resp. right) ideals $L_{U,P}$ (resp. $R_{U,P}$) contained in I. (Note that, in view of 2.12.3, minimal left and right ideals are composed of functions which are equal a.e. to t.p.s.; orthogonality of two such functions is to be interpreted in the sense of the scalar product on $L^2(G)$.)

Proof. It suffices to deal with the left-handed case. Further, one may and will assume that $I \neq \{\underline{0}\}$. By 2.9.2 and 2.9.3, any $f \in I$ is the limit of finite sums of mutually orthogonal t.p.s.

$$f_U : x \mapsto d(U). \mathrm{Tr}[\hat{f}(U)U(x)^*] = d(U)\chi_U * f(x) ,$$

each of which belongs to I (since I is a left ideal). It is thus enough to show that each f_U is a linear combination of functions $g_i \in L_{U,P_i}$, where the P_i are mutually orthogonal one-dimensional projectors on $\mathscr{H}_U$ such that $L_{U,P_i} \subseteq E * f_U \subseteq I$. However, writing $A = \hat{f}(U) = \hat{f}_U(U)$, $L_{U,P} \subseteq E * f_U$ if and only if $P = XA$ for some $X \in \mathrm{End}\,(\mathscr{H}_U)$. On the other hand, Lemma A.3.7 of Appendix A affirms that A is a sum $\Sigma_i AP_i$, where the P_i are mutually orthogonal one dimensional projectors on $\mathscr{H}_U$, each of the form $P_i = X_i A$ for some $X_i \in \mathrm{End}\,(\mathscr{H}_U)$. It then suffices to take $g_i = f_U * \phi_{U,P_i} = f * \phi_{U,P_i}$

2.12.5. Turning to consider (two-sided) ideals, denote by E_U the set of $f \in E$ such that $\hat{f}(V) = 0$ for every $V \in \hat{G}$ different from U. It is almost evident that $E_U \neq \{\underline{0}\}$ is an ideal; that E_U is stable under the mapping $f \mapsto \tilde{f}$; and that

$$E_U = \chi_U * E = E * \chi_U = \chi_U * E * \chi_U = E * \chi_U * E .$$

Moreover, if U and V are distinct elements of $\hat{G}$,

$$E_U \cap E_V = \{\underline{0}\} , \tag{2.12.1}$$

$$f * \tilde{g} = \underline{0} \text{ if } f \in E_U,\ g \in E_V , \tag{2.12.2}$$

the second of which says that the E_U are mutually orthogonal in the $L^2(G)$ sense.

2.12.6. **Theorem.** A two-sided ideal in E is minimal if and only if it is of the form E_U.

Proof. Let I be minimal two-sided. If $f \in I$ and $U \in \hat{G}$ be chosen so that $\hat{f}(U) \neq 0$, Lemma A. 3. 5 of Appendix A shows that $I \supseteq E_U$. Minimality of I implies that $I = E_U$.

On the other hand, to show that every E_U is minimal two-sided, note that if I is two-sided and $\subseteq E_U$, the hypothesis that $I \neq \{0\}$ leads (Lemma A. 3. 5 of Appendix A again) to the conclusion that $I = E_U$. So E_U is indeed minimal two-sided.

2.12.7. **Theorem.** Any (two-sided) ideal I in E is contained in the closure in E of the direct sum of those E_U contained in I.

Proof. Let $f \in I$. Using the notation introduced in the proof of 2.12.4, 2.9.2 and 2.9.3 show that f is the limit in E of finite sums Σf_U involving only those $U \in \hat{G}$ for which $\hat{f}(U) \neq 0$. On the other hand, Lemma A. 3. 5 of Appendix A shows that the subspace of E generated by $E * f * E$, which is contained in I, contains E_U if (and only if) $\hat{f}(U) \neq 0$.

On combining 2.12.4 and 2.12.7 there appears

2.12.8. **Theorem.** Every closed left (resp. right, two-sided) ideal I in E is the closure in E of the direct sum of certain mutually orthogonal minimal left (resp. right, two-sided) ideals $L_{U,P}$ (resp. $R_{U,P}$, E_U) contained in I.

2.12.9. The main point about 2.12.8 is that it contains an assurance that a closed ideal I in E (left, right or two-sided) is completely determined by its 'components' $I \cap E_U$, one for each $U \in \hat{G}$, each $I \cap E_U$ being an ideal (of the same type) in the f. d. algebra E_U. If this be coupled with the fact that the Fourier transform sets up an isomorphism between E_U and End $(\mathscr{H}_U)$, one sees that in the end everything is reduced to describing the ideals in End $(\mathscr{H}_U)$ ... an almost purely algebraic problem. In this way, 2.12.8 leads to the following more concrete descriptions.

(i) To every closed left (resp. right) ideal I in E corresponds a family $(P_U)_{U \in \hat{G}}$, where P_U is an orthogonal projector on $\mathcal{H}_U$, such that

$$I = \{f \in E : \hat{f}(U)P_U = 0 \text{ for every } U \in \hat{G}\}$$
$$(\text{resp. } I = \{f \in E : P_U\hat{f}(U) = 0 \text{ for every } U \in \hat{G}\}) ;$$

and conversely.

(Compare this with 2.10.4 and 2.10.5; see also Hewitt and Ross [1], (38.13).)

The case of closed two-sided ideals is especially simple, being covered by the following.

(ii) To every closed two-sided ideal I in E corresponds a subset S of $\hat{G}$ such that

$$I = \{f \in E : \hat{f}(U) = 0 \text{ for every } U \in S\}.$$

The proof of (i) is reduced to that of the appropriate assertion about left (resp. right) ideals in End ($\mathcal{H}$), where $\mathcal{H}$ denotes a f. d. Hilbert space.

For the case of left ideals $\mathcal{I}$ in End ($\mathcal{H}$), one has merely to prove the existence of a linear subspace $\mathcal{M}_0$ of $\mathcal{H}$ such that

$$\mathcal{I} = \{X \in \text{End}(\mathcal{H}) : X(\mathcal{M}_0) = \{0\}\} ;$$

for this is equivalent to saying that

$$\mathcal{I} = \{X \in \text{End}(\mathcal{H}) : XP_0 = 0\} ,$$

where P_0 is the orthogonal projection of $\mathcal{H}$ onto $\mathcal{M}_0$. A detailed proof of this appears in Hewitt and Ross [1], (38.11).

The case of right ideals is easily derived from this by passage to the adjoint. If $\mathcal{I}$ is a right ideal in End ($\mathcal{H}$), then $\mathcal{I}^*$ is a left ideal therein; so, by what has just been indicated $\mathcal{I}^* = \{X : XP_0 = 0\}$ for a suitable P_0; hence $\mathcal{I} = (\mathcal{I}^*)^* = \{Y \in \text{End}(\mathcal{H}) : Y^*P_0 = 0\} = \{Y : P_0Y = 0\}$, which is the desired conclusion. (Recall that

$(P_0 Y)^* = Y^* P_0$ since P_0 is s. a.; and that $A = 0$ if and only if $A^* = 0$.)

The two-sided case can be derived by combining the left- and right-cases, or more directly from Appendix A. 3. 5 (which goes to show that the only two-sided ideals in End $(\mathscr{H})$ are $\{0\}$ and End $(\mathscr{H})$ itself.

2.12.10 **Maximal closed ideals.** Results like those in 2.12 8 and 2.12.9 are often reformulated and complemented by statements involving maximal, rather than minimal, ideals. We describe briefly some such results, seeking simplicity by considering only closed (two-sided) ideals in E.

A closed ideal M is termed maximal if it is proper (i. e., different from E) and is contained properly in no closed ideal different from E.

It is then possible to prove the following:

(i) The maximal closed ideals in E are precisely the sets

$$M_U = \{f \in E : \hat{f}(U) = 0\}, \qquad (2.12.3)$$

one corresponding to each $U \in \hat{G}$.

(ii) Every closed ideal I in E is the intersection of those maximal closed ideals which contain it:

$$I = \bigcap \{M_U : U \in \hat{G},\ M_U \supseteq I\}. \qquad (2.12.4)$$

These assertions may be inferred from 2.12.9(ii), or proved directly by using the Hahn-Banach theorem combined with the substance of 2.9.3 and 2.9.6. In (ii) one has effectively the extension of (2.10.7) to the non-Abelian case.

For some similar results applying to one-sided ideals, see Hewitt and Ross [1], §38.

2.12.11. **Closed invariant subspaces.** Turning to closed invariant subspaces, it is first of all almost evident that $L_{U,P}$ (resp. $R_{U,P}$, E_U) is a closed left (resp. right, two-sided) invariant subspace of E. It is true, but not nearly so obvious, that the following analogue of 2.12.3 and 2.12.6 obtains: the minimal left (resp. right, two-sided) invariant subspaces of E are precisely the $L_{U,P}$ (resp. $R_{U,P}$, E_U).

2.12.12. **Theorem.** Every closed left (resp. right, two-sided) invariant subspace in E is the closure in E of the direct sum of certain mutually orthogonal minimal left (resp. right, two-sided) invariant subspaces $L_{U,P}$ (resp. $R_{U,P}$, E_U), each of which is contained in the given subspace.

Proof. This is a direct consequence of 2.12.1, 2.12.8 and 2.12.11.

2.12.13. There is no analogue of 2.12.4 or 2.12.7 applying to not-necessarily-closed invariant subspaces. There are in general non-closed invariant subspaces, which are 'large' in the sense of being everywhere dense in E, and which nonetheless contain no non-zero t. p. s. (and therefore contain no minimal invariant subspace, and no closed invariant subspace other than $\{0\}$); see Exercise 2.12.18.

2.12.14. It is now possible to verify a statement made toward the end of 2.1.10, namely: every $U \in \hat{G}$ is unitarily equivalent to a representation of G of the form $x \mapsto L_x | M$, where M is some minimal left invariant subspace of $L^2(G)$.

In fact, take any $\underset{\sim}{e} \in \mathscr{H}_U$ satisfying $\|\underset{\sim}{e}\| = 1$ and let P be the orthogonal projector of $\mathscr{H}_U$ onto the subspace generated by $\underset{\sim}{e}$. Let

$$\phi(x) = \phi_{U,P}(x) = (\underset{\sim}{e} | U(x)\underset{\sim}{e})$$

and take M to be the minimal left invariant subspace $L_{U,P}$ of $L^2(G)$. M consists precisely of all linear combinations of left translates of ϕ, i. e., of functions

$$f_{\underset{\sim}{v}} : x \mapsto (\underset{\sim}{v} | U(x)\underset{\sim}{e})$$

where $\underset{\sim}{v} \in \mathscr{H}_U$. Since U is irreducible, the $U(x)\underset{\sim}{e}$ generate the whole of $\mathscr{H}_U$, and so $f_{\underset{\sim}{v}}$ determines $\underset{\sim}{v}$ uniquely. There is therefore a linear map $J : \underset{\sim}{v} \mapsto d(U)^{\frac{1}{2}} f_{\underset{\sim}{v}}$ of $\mathscr{H}_U$ onto M. On using (2.6.3), one sees that J preserves scalar products. Moreover, $L_x J = JU(x)$ for every $x \in G$, and so

$$L_x | M = JU(x)J^{-1}$$

which exhibits the alleged unitary equivalence of U and $x \mapsto L_x | M$.

2.12.15. The Abelian case. There is now no distinction between $L_{U,P}$, $R_{U,P}$ and E_U: all three are identical with the one-dimensional subspace of E generated by the (bounded continuous multiplicative) character χ of U. The maximal closed ideals in E are precisely the sets

$$M_\chi = \{f \in E : \hat{f}(\chi) = 0\},$$

where $\chi \in \Gamma$. Every closed ideal I in E (alternatively: every closed invariant subspace I of E) consists precisely of those $f \in E$ such that $\hat{f}(\chi) = 0$ for every $\chi \in Z_I$, where $Z_I \subseteq \Gamma$ is the set of common zeros of Fourier transforms of elements of I. (This is virtually a reformulation of (2.10.7).)

2.12.16. Exercise. Prove that a subset I of E is a closed ideal in E of a given type (i.e., left, right or two-sided) if and only if it is a closed invariant subspace of E of the same type.

2.12.17. Exercise. Prove that the smallest left ideal in E containing $f \in E$ is

$$\{\lambda f + g * f : \lambda \in \mathbf{C},\ g \in E\},$$

and that the smallest closed left ideal in E containing f is the closure of $E * f$.

2.12.18. Exercise. Take $G = \mathbf{T}$ and $f \in C(\mathbf{T})$ defined by

$$f(e^{it}) = \Sigma_{n \in \mathbf{Z}}\, c_n e^{int}$$

where $\Sigma_{n \in Z} |c_n| < \infty$ and $c_n \neq 0$ for $n > n_0$. Show that the invariant subspace of $C(\mathbf{T})$ generated by f contains no non-zero t.p.

2.13. Spectral synthesis problems

A stage has now been reached at which the reader should experience no difficulty in appreciating a class of problems of current interest in abstract harmonic analysis, namely, the so-called spectral synthesis problems. This section undertakes to explain briefly the nature of these problems and the connections between them and the results already covered in earlier sections of these notes.

2.13.1. The original spectral synthesis problem, which was introduced by Beurling a little more than thirty years ago, concerned the approximation of any given $f \in L^{\infty}(\mathbf{R})$ by linear combinations of those (bounded continuous multiplicative) characters of $\mathbf{R}$ which belong to the closed span of translates of f. The topology envisaged by Beurling was the 'narrow' topology mentioned in 2.11.4, but most subsequent work uses the weak topology on $L^{\infty}(\mathbf{R})$ as the dual of $L^1(\mathbf{R})$.

When approximation is understood in the latter sense, Beurling's problem can (by general duality theory) be formulated as that of expressing a general closed ideal I in the convolution algebra $L^1(\mathbf{R})$ as the intersection of maximal ideals in $L^1(\mathbf{R})$.

This original problem has since given rise to the discussion of a wide variety of somewhat similar problems, all described (at times a little bewilderingly) as <u>spectral</u> (or <u>harmonic</u>) <u>synthesis problems</u>. The name is suggested by the fact that some problems of this type (including the original one) are indeed concerned with synthesising a given function from its harmonic constituents, though other problems to which the same description is applied do not naturally conform to this type (and might equally well be referred to as problems of analysis rather than synthesis). The appropriate concepts and language vary from problem to problem: sometimes minimal ideals are appropriate, sometimes maximal ideals, and sometimes invariant subspaces. (Broadly speaking, minimal ideals are rather rare birds and the concept of maximal ideal is more generally useful in this context. The convolution algebras E discussed in 2.10 and 2.12 above offer an unusually wide freedom of choice, this being mainly due to the fact that the group in question is assumed to be compact.)

A substantial variety (but by no means all) of the interesting spectral synthesis problems can be formulated in the following way, at least if one allows for the interchange of isomorphic algebras.

2.13.2. Suppose given a set S, a family $(A_s)_{s \in S}$ of complex algebras, and a subalgebra A of the product algebra $\Pi_{s \in S} A_s$. The elements of A are thus functions f with domain S such that $f(s) \in A_s$ for every $s \in S$, and the algebraic operations in A are 'pointwise' or 'coordinatewise'. We assume that each A_s is topologised in such a way that $\{0_s\}$ is a closed subset of A_s, where 0_s denotes the zero element of A_s; and that A is topologised in such a way that $f \mapsto f(s)$ is continuous from A into A_s for every $s \in S$.

There are various species of ideals in A (left ideals, closed left ideals, right ideals, and so on), together with maximal ideals of each species, but this is no place to attempt even a sketch of the associated questions. Suffice it to mention that in the commutative case one usually attempts to work with those maximal ideals which are modular (or regular), these being the maximal ideals most closely linked with kernels of non-zero homomorphisms of A into the complex field; see Exercise 2.13.4. [A left (resp. right) ideal I in A is said to be modular if there exists in A an element u which is a right (resp. left) identity modulo I, i.e., an element u which satisfies $fu - f \in I$ (resp. $uf - f \in I$) for every $f \in A$; a (two-sided) ideal is modular if there exists in A an element which is both a right and a left identity modulo I.]

In this situation, the so-called spectral synthesis problem for A amounts to determining which ideals in A are intersections of maximal ideals of various species.

The most obvious ideals in A are the closed ideals

$$I(X) = \{f \in A : f(s) = 0_s \text{ for every } s \in X\}, \qquad (2.13.1)$$

where X denotes a subset of S. Reciprocally, to every ideal I in A corresponds the subset

$$Z(I) = \{s \in S : f(s) = 0_s \text{ for every } f \in I\} \qquad (2.13.2)$$

of S. It is simple to verify the relations

$$\left.\begin{array}{l} X \subseteq Z(I(X)), \quad I \subseteq I(Z(I)), \\ I(X) \subseteq I(Y) \text{ if } Y \subseteq X \subseteq S. \end{array}\right\} \qquad (2.13.3)$$

A moment's thought makes it very plausible that the maximal closed ideals in A will be the sets

$$M_s = I(\{s\}) = \{f \in A : f(s) = 0_s\} , \qquad (2.13.4)$$

one corresponding to each point s of S. In many cases of interest, this conjecture proves to be true. Usually, each M_s is modular; they may also be maximal ideals. Irrespective of the precise properties of the M_s, the spectral synthesis problem is often interpreted to be that of determining which closed ideals I in A are of the form

$$I(X) = \bigcap \{M(s) : s \in X\}$$

for a suitable subset X of S. Now, if $I = I(X)$ holds for any subset X of S, (2.13.3) shows that $X = Z(I)$ is one such set. Thus the problem boils down to deciding which closed ideals I in A satisfy

$$I = I(Z(I)) . \qquad (2.13.5)$$

It turns out that in some cases (2.13.5) holds for every closed ideal I in A. In certain other cases it will be evident from the outset that there are closed ideals I in A for which (2.13.5) is false. In still other cases, considerable effort and ingenuity may be needed to produce (or even merely prove the existence of) closed ideals I in A for which (2.13.5) is false. In this last case, various interesting problems may present themselves. For example:

(i) Can one specify conditions upon $Z(I)$ sufficient (or perhaps even necessary and sufficient) to ensure the truth of (2.13.5)?

(ii) Given an element f of A such that $f \in I(Z(I))$, can one specify additional conditions on f sufficient (or maybe necessary and sufficient) to ensure that $f \in I$?

Many problems of this sort are very delicate; while numerous sufficient conditions are known, the question of necessary and sufficient conditions is almost always unsolved.

Some specific examples are worth looking at briefly.

2.13.3. Some examples

(i) Here S is a compact Hausdorff space, A_s is the complex field for every $s \in S$, and A is a subalgebra of the algebra $C(S)$ of continuous complex-valued functions on S. It is always possible to endow

A with the sup norm ($\|f\| = \sup\{|f(s)| : s \in S\}$), though other norms may be used for certain A's.

Suppose first that $A = C(S)$ itself. It can then be shown without much trouble that the maximal ideals in A are exactly those given by (2.13.4). (All ideals in A are modular because A has an identity element, the constant function $\underline{1}$.) It can also be shown that (2.13.5) is true for every closed ideal I in A. Problems (i) and (ii) in 2.13.2 do not arise.

However, the situation may be otherwise if one takes for A a proper subalgebra of C(S); see (iii) below.

(ii) Let G be a compact group, $S = \hat{G}$, $A_U = \text{End}(\mathscr{H}_U)$ for $U \in \hat{G}$, and let E be as specified in 2.12.10 above. This situation can be fitted into the scheme described in 2.13.2 by taking for A the set of Fourier transforms $\hat{f}$ with $f \in E$.

In this case 2.12.10(ii) affirms that once again (2.13.5) is valid for every closed ideal I in A. (It was seen in 2.12.10(i) that the maximal closed ideals follow the pattern described in (2.13.4).)

An equivalent version of the result appears in 2.12.8, which is genuinely a spectral synthesis assertion.

Once again, the relatively subtle problems (i) and (ii) in 2.13.2 do not arise.

(iii) Turning to what will prove to be a more interesting example, refer back to (i) above and take therein $S = G$, a compact group, and $A = A(G)$ as defined in 2.9.8. The maximal ideals in A can be shown to be given by (2.13.4). (The proof is relatively easy when G is Abelian (see Exercise 2.13.4 below); the contrary case is more difficult and is dealt with in Eymard [1].) However, even for familiar choices of G (the circle group **T**, for example), (2.13.5) is <u>not</u> true for every closed ideal I in A(G); questions (i) and (ii) of 2.13.2 arise and have received a great deal of attention.

Concerning (i) of 2.13.2, it is known that (2.13.5) holds for every closed ideal I in A(G) such that the frontier of Z(I) contains no non-void perfect subset. (For Abelian G, see Edwards [3], 12.11.4-12.11.6 and Exercise 12.52; for general G, the result is a consequence of Theorem (4.19) of Eymard [1]; see also Hewitt and Ross [1], (39.24).

As regards (ii) of 2.13.2, if $G = \mathbf{T}$, it is the case that any closed ideal I in A(G) contains every $f \in A(G)$ which vanishes 'sufficiently strongly' on Z(I) (see Edwards [3], 12.11.4 and 13.5.5)... but no interpretation of 'sufficiently strongly' is known which is necessary and sufficient. To this extent, the problem remains open (and presumably very difficult).

(iv) To get back to a case which covers instances 'isomorphic' to Beurling's original problem (see 2.13.1), let G be a locally compact Abelian group; $G = \mathbf{R}^n$ is pretty typical. Let Γ denote the group dual to G; see 2.5.4. Take $S = \Gamma$ and A_χ = the complex field for every $\chi \in \Gamma$. Finally, take for A the set of Fourier transforms $\hat{f}$ where $f \in L^1(G)$; for good reasons, this A is usually denoted by $A(\Gamma)$.

The state of affairs here turns out to be much as is described in (iii) immediately above for the case of A(G) with G an infinite compact group; see Edwards [3], 11.2.3, 12.11.4 and Exercise 12.53 for a few more details. (There is, however, the difference that Γ is in general neither discrete nor compact.)

$A(\Gamma)$ is, by its definition, isomorphic with $L^1(G)$ and the problem is that of expressing a general closed ideal in $L^1(G)$ as an intersection of modular maximal ideals in $L^1(G)$ (cf. the second paragraph in 2.13.1).

This is one of those cases in which it took time and effort to decide whether or not (2.13.5) holds for every closed ideal I. The first examples showing that (2.13.5) need <u>not</u> hold were produced by Schwartz in 1948 and applied only to the cases $G = \mathbf{R}^n$ with $n \geq 3$. By this time, the problem had been 'in the air' for about a decade; and a further decade was to elapse before Rudin, Kahane and Malliavin coped with the case of a general non-compact locally compact Abelian group G. Schwartz's work actually produces closed ideals I for which (2.13.5) false, but the Rudin-Kahane-Malliavin technique suffices only to prove the <u>existence</u> of such ideals I (the situation being analogous to that described in Remark (iii) in 2.15.6).

For many more details concerning these examples of the failure of spectral synthesis, see Hewitt and Ross [1], Chapter X. Most especially we refer the reader to the account given loc. cit. §42 of a novel approach devised by Varopoulos, which amounts to a strikingly original importation into harmonic analysis of ideas relating to tensor products of Banach spaces and algebras. (The functional analytic theory of tensor products goes back for 20 years or more; in the context of topological linear spaces it underwent very broad development by Grothendieck (c. 1955), but its use in harmonic analysis appears to have been an especially happy and fruitful turn in events (c. 1964) relating to the Rudin-Kahane-Malliavin technique.)

2.13.4. Exercise. Assume the following two basic facts:

(i) If A is a commutative complex Banach algebra, every modular maximal ideal in A is the kernel of a continuous homomorphism τ of A onto **C**. (See Hewitt and Ross [1], Appendix (C.17).)

(ii) The Pontryagin duality law, as stated in 2.5.4(b) above.

Let G be a compact Abelian group. Show that every maximal ideal in A(G) conforms to the formula (2.13.4) for some $s \in G$.

2.14. The Hausdorff-Young theorem

In this section we return to some more concrete aspects of Fourier transforms, the topic being one which might be made to follow on from the substance of 2.7. In brief, we shall consider some important partial extensions of the Parseval formula.

In order to explain what we are up to, it is undoubtedly best to begin by looking at the Abelian case.

2.14.1. Preliminary discussion of the Abelian case. Throughout this subsection, G will denote a compact Abelian group with dual group Γ; see 2.5.4.

If ψ is a complex-valued function on Γ and $p \in [1, \infty]$ it is standard practice to write

$$\|\psi\|_p = \{\Sigma_{\chi \in \Gamma} |\psi(\chi)|^p\}^{1/p} \quad \text{if } p \neq \infty$$

and

$$\|\psi\|_\infty = \sup_{\chi \in \Gamma} |\psi(\chi)| \quad ,$$

allowing ∞ as a possible value of $\|\psi\|_p$. It is also standard to write $l^p(\Gamma)$ for the linear space of complex-valued functions ψ on Γ for which $\|\psi\|_p < \infty$. Then the restriction of $\psi \mapsto \|\psi\|_p$ to $l^p(\Gamma)$ is a norm on $l^p(\Gamma)$ relative to which $l^p(\Gamma)$ is a Banach space. In addition, we shall write $l_c(\Gamma)$ for the set of complex-valued functions ψ on Γ having finite supports, noting that $l_c(\Gamma)$ is a linear subspace of $l^p(\Gamma)$ for every p. To complete the range on display, $c_0(\Gamma)$ is defined to be the closure in $l^\infty(\Gamma)$ of $l_c(\Gamma)$. It is easy to check that a complex-valued function ψ on Γ belongs to $c_0(\Gamma)$ if and only if

$$\lim_{\chi \in \Gamma, \chi \to \infty} \psi(\chi) = 0 ,$$

the limit being interpreted in the fashion explained in 1.2.5 above. It is worth noting that

$$l_c(\Gamma) \subseteq l^p(\Gamma) \subseteq c_0(\Gamma) \subseteq l^\infty(\Gamma) \text{ if } p \in [1, \infty),$$

the inclusions being strict whenever Γ is infinite; and that

$$\|\psi\|_q \leq \|\psi\|_p \text{ if } q \geq p .$$

(With the obvious conventions, this last holds even in cases where either or both of $\|\psi\|_q$ and $\|\psi\|_p$ are ∞.)

If $f \in L^1(G)$ (or even to $M(G)$), $\hat{f}$ is a complex-valued function on Γ. The inequality (2.3.2) is equivalent to the formula

$$\|\hat{f}\|_\infty \leq \|f\|_1 ,$$

and 2.7.4 asserts exactly that

$$\hat{f} \in c_0(\Gamma) .$$

The substance of 2.7 affirms that $f \mapsto \hat{f}$ maps $L^2(G)$ onto $l^2(\Gamma)$ and the Parseval formula says that this mapping is an isometry:

$$\|\hat{f}\|_1 = \|f\|_2 .$$

The original Hausdorff-Young theorem applied to the case $G = \mathbf{T}$ and asserts that then if $p \in [1, 2]$, $f \mapsto \hat{f}$ maps $L^p(G)$ into $l^{p'}(\Gamma)$, where $1/p + 1/p' = 1$ (conventionally $p' = \infty$ when $p = 1$), and that

$$\|\hat{f}\|_{p'} \leq \|f\|_p . \qquad (2.14.1)$$

For a discussion of this result (by methods which extend without trouble to any compact Abelian G), see Edwards [3], Section 13. 5.

The cases $p = 1$ and $p = 2$ tell us nothing new; but for $1 < p < 2$, the fact that $\hat{f} \in l^{p'}(\Gamma)$ whenever $f \in L^p(G)$, is a definite advance.

When compared with the case $p = 2$, the Hausdorff-Young theorem constitutes only a partial replacement (albeit a very useful one): only in trivial cases does $f \mapsto \hat{f}$ map $L^1(G)$ onto $c_0(\Gamma)$; and, if $1 < p < 2$, only in trivial cases is it true that $f \mapsto \hat{f}$ maps $L^p(G)$ onto $l^{p'}(\Gamma)$ or that one has equality in (2. 14. 1). See 2. 14. 3 and 2. 14. 6 below.

We now wish to pass on to the description of the Hausdorff-Young theorem as it applies to general compact groups.

The first step is the introduction of suitable analogues of the spaces $l^p(\Gamma)$ and their norms $\|.\|_p$, bearing in mind that now complex-valued functions on Γ will have to be replaced by functions Ψ on $\hat{G}$ such that $\Psi(U) \in \text{End}\ (\mathscr{H}_U)$ for every $U \in \hat{G}$. This is by no means a trivial matter, and we must assume that the reader will at this point refer to Appendix A. 4. Subsections 2. 14. 2 and 2. 14. 3 are concerned with this topic.

The other major preliminary is a convexity theorem about bilinear functionals; this will be described in 2. 14. 4.

2. 14. 2. The spaces $E^p(\hat{G})$. In terms of the norms $\|.\|_{\phi_p}$ described in Appendix A. 4. 2, we can define certain linear subspaces of

$$E(\hat{G}) = \Pi_{U \in \hat{G}}\ \text{End}\ (\mathscr{H}_U)$$

which are easily seen to be generalisations, to the non-Abelian case, of the spaces $l^p(\Gamma)$ figuring in 2. 14. 1, and which prove to be just right for the statement of the Hausdorff-Young theorem valid for all compact groups.

Thus, for $\Psi \in E(\hat{G})$, we define

$$
\left.\begin{aligned}
\|\Psi\|_p &= \{\Sigma_{U\in\hat{G}}\, d(U)\,\|\Psi(U)\|_{\phi_p}^p\}^{1/p} \quad \text{if } p\in[1,\infty)\,, \\
\|\Psi\|_\infty &= \sup_{U\in\hat{G}}\,\|\Psi(U)\|_{\phi_\infty}\,,
\end{aligned}\right\} \quad (2.14.2)
$$

either or both of which may be ∞. We then write $E^p(\hat{G})$ for the linear space of $\Psi \in E(\hat{G})$ for which $\|\Psi\|_p < \infty$; $E_c(\hat{G})$ for the linear space of $\Psi \in E(G)$ for which the set $\{U \in \hat{G} : \Psi(U) \neq 0\}$ is finite; and $E_0(\hat{G})$ for the closure in $E^\infty(\hat{G})$ of $E_c(\hat{G})$. This last is also the set of $\Psi \in E(\hat{G})$ such that

$$
\lim_{U\in\hat{G},\, U\to\infty} \|\Psi(U)\|_\infty = 0\,, \quad (2.14.3)
$$

where the limit is interpreted as explained in Exercise 1.2.5.

It can be verified that $E^p(\hat{G})$ is a Banach space when taken with the norm equal to the restriction to $E^p(\hat{G})$ of $\Psi \mapsto \|\Psi\|_p$; and that $\|\Psi^*\|_p = \|\Psi\|_p$ for every $\Psi \in E(\hat{G})$, where, of course, $\Psi^* : U \mapsto (\Psi(U))^*$.

2.14.3. The results of 2.3 and the orthogonality relations show that the Fourier transformation $FT : f \mapsto \hat{f}$ maps $M(G)$ into $E^\infty(\hat{G})$ and $T(G)$ into $E_c(\hat{G})$. On the other hand, 2.7.4 asserts precisely that FT maps $L^1(G)$ into $E_0(\hat{G})$.

It can be shown that the image under FT of $L^1(G)$ is a dense subspace of $E_0(\hat{G})$; and that this image is the whole of $E_0(\hat{G})$ if and only if G is finite. (See Hewitt and Ross [1], (28.40) and (37.4); for the Abelian case see also Edwards [3], 2.3.9 and Exercise 2.14.10 below.)

$E^2(\hat{G})$ is rather special. To begin with, it is a Hilbert space relative to the scalar product

$$
(\Psi_1 | \Psi_2) = \Sigma_{U\in\hat{G}}\, d(U).\,\mathrm{Tr}[\Psi_1(U)\Psi_2(U)^*]\,,
$$

the associated norm being $\|.\|_2$, as defined in (2.14.2). Moreover the Parseval formula shows that FT is an isometric linear map of $L^2(G)$ onto the whole of $E^2(\hat{G})$.

It is also worth noting that the space $A(G)$ defined in 2.9.8 is precisely the set of $f \in C(G)$ such that $\|\hat{f}\|_1 < \infty$, and that $\|f\|_{A(G)} = \|\hat{f}\|_1$ for every $f \in A(G)$. In other words, FT is a linear isometry of $A(G)$ onto $E^1(\hat{G})$.

2.14.4. **The convexity theorem.** A discussion of some theorems of this type, applying to the case of linear operators acting on suitable space of complex-valued functions on some measure space, will be found in Edwards [3], Chapter 13. These theorems are admirably suited to the proof of the Hausdorff-Young theorem for Abelian groups G; see loc. cit. Section 13.5. For the non-Abelian case, however, it is necessary to modify the convexity theorems so as to apply to operators acting on suitable spaces of vector-valued functions. A modification of this sort is established in Hewitt and Ross [1], (E.18), to which we must refer the reader for details. In either case it is largely a matter of convenience whether one chooses to talk about linear operators or about bilinear functionals (one may, indeed, increase the generality and talk about multilinear operators). We will here merely cite a special version of the convexity theorem, expressed in terms of bilinear functionals and including enough generality to satisfy our immediate needs (with a little, but not much, to spare). The statement follows.

Let B be a complex-valued bilinear functional on $T(G) \times E_c(\hat{G})$ such that, for certain p_i, $q_i \in [1, \infty]$ and $M_i \in (0, \infty)$ $(i = 1, 2)$, one has

$$|B(f, \Psi)| \leq M_1 \|f\|_{p_1} \|\Psi\|_{p_2}, \tag{2.14.4}$$

$$|B(f, \Psi)| \leq M_2 \|f\|_{q_1} \|\Psi\|_{q_2} \tag{2.14.5}$$

for every $(f, \Psi) \in T(G) \times E_c(\hat{G})$. Suppose that $t \in [0, 1]$ and that

$$1/r_1 = (1-t)/p_1 + t/q_1, \quad 1/r_2 = (1-t)/p_2 + t/q_2 . \tag{2.14.6}$$

Then

$$|B(f, \Psi)| \leq M \|f\|_{r_1} \|\Psi\|_{r_2} \tag{2.14.7}$$

for every $(f, \Psi) \in T(G) \times E_c(\hat{G})$, where $M \leq M_1^{1-t} M_2^t$.

In the terminology used by Hewitt and Ross, loc. cit., the hypotheses (2.14.4) and (2.14.5) assert precisely that B is of type $(p_1, p_2; 1)$ and of type $(q_1, q_2; 1)$, respectively; and the conclusion (2.14.7) is that it is then of type $(r_1, r_2; 1)$, whenever the r's are convex combinations of the p's and q's of the sort prescribed in (2.14.6). (This is the genesis of the term 'convexity theorem'.)

All the necessary armament has now been assembled.

2.14.5. Theorem (Hausdorff-Young). If $1 \le p \le 2$ and $f \in L^p(G)$, then $\hat{f} \in E^{p'}(\hat{G})$ and

$$\|\hat{f}\|_{p'} \le \|f\|_p . \tag{2.14.1}$$

Proof. We shall apply the convexity theorem to the bilinear functional B defined by

$$B(f, \Psi) = \Sigma_{U \in \hat{G}} d(U). \operatorname{Tr} \hat{f}(U)\Psi(U) ;$$

note that the summand is non-zero for at most a finite set of $U \in \hat{G}$ (depending upon (f, Ψ)). We proceed to verify that B satisfies certain 'type inequalities'.

By equation (A.4.7) of the Appendix and (2.3.2) above,

$$\begin{aligned} |\operatorname{Tr} \hat{f}(U)\Psi(U)| &\le \|\hat{f}(U)\|_{\phi_\infty} \|\Psi(U)\|_{\phi_1} \\ &\le \|f\|_1 \|\Psi(U)\|_{\phi_1} ; \end{aligned} \tag{2.14.8}$$

and, by (A.4.7) again,

$$|\operatorname{Tr} \hat{f}(U)\Psi(U)| \le \|\hat{f}(U)\|_{\phi_2} \|\Psi(U)\|_{\phi_2} . \tag{2.14.9}$$

By (2.14.8),

$$|B(f, \Psi)| \le \|f\|_1 \Sigma_{U \in \hat{G}} d(U) \|\Psi(U)\|_{\phi_1} = \|f\|_1 \|\Psi\|_1 ; \tag{2.14.10}$$

and, by (2.14.9) and the Parseval formula,

$$\begin{aligned} |B(f, \Psi)| &\le \Sigma_{U \in \hat{G}} d(U) \|\hat{f}(U)\|_{\phi_2} \|\Psi(U)\|_{\phi_2} \\ &\le \{\Sigma_{U \in \hat{G}} d(U) \|\hat{f}(U)\|^2_{\phi_2}\}^{\frac{1}{2}} \{\Sigma_{U \in \hat{G}} d(U) \|\Psi(U)\|^2_{\phi_2}\}^{\frac{1}{2}} \\ &= \|f\|_2 \|\Psi\|_2 . \end{aligned} \tag{2.14.11}$$

Of these, (2.14.10) says that B is of type (1, 1; 1), and (2.14.11) that it is of type (2, 2; 1). The convexity theorem of 2.14.4 accordingly affirms that B is also of type (p, p; 1) whenever $1 \le p \le 2$, and that

$$|B(f, \Psi)| \le \|f\|_p \|\Psi\|_p \qquad (2.14.12)$$

for such values of p. The rest of the proof consists of deriving (2.14.1) from (2.14.12).

To do this, we call upon (A.4.6) of the Appendix: this shows that, for given f and given U, $\Omega(U) \in \mathrm{End}(\mathscr{H}_U)$ may be selected so that

$$\|\Omega(U)\|_{\phi_p} = 1 \text{ and } \|\hat{f}(U)\|_{\phi_{p'}} = \mathrm{Tr}\, \hat{f}(U)\Omega(U) ; \qquad (2.14.13)$$

plainly, if $\hat{f}(U) = 0$, one may take $\Omega(U) = 0$. So, given $f \in T(G)$, Ω may be selected from $E_c(\hat{G})$ so that (2.14.13) holds for every $U \in \hat{G}$. Now apply (2.14.12), taking

$$\Psi : U \mapsto \|\hat{f}(U)\|_{\phi_{p'}}^{p'-1} \Omega(U) ,$$

noting that $\Psi \in E_c(\hat{G})$, that

$$\|\Psi(U)\|_{\phi_p}^{p} = \|\hat{f}(U)\|_{\phi_{p'}}^{p'} ,$$

and so that

$$\|\Psi\|_p^p = \|\hat{f}\|_{p'}^{p'} .$$

It then appears that

$$\|\hat{f}\|_{p'}^{p'} \le \|f\|_p \|\hat{f}\|_{p'}^{p'/p} ,$$

which, since $p' - p'/p = 1$, is equivalent to (2.14.1).

Thus, (2.14.1) is established for every $f \in T(G)$. We leave to the reader the task of deriving (2.14.1) for every $f \in L^p(G)$; see Exercise 2.14.7.

2.14.6. There are various ways in which one might hope to improve the Hausdorff-Young theorem... by replacing the inequality by an equality, for example, or by enlarging the allowed set of values of p.

However, most hopes of this sort prove to be ill-founded. For the Abelian case, see Exercise 2.14.9 below and Edwards [3], 13.5.3, 13.5.4, 14.4 and 15.4; for the general case, see Hewitt and Ross [1], (37.19). It would seem that the stated version is about the best one can hope for in general.

There is, however, a 'dual' version of Hausdorff-Young's inequality which applies to functions on $\hat{G}$. In 2.14.3 we have noted that FT maps $L^2(G)$ isometrically onto $E^2(G)$; let FT^{-1} denote the inverse map. If $1 \le p \le 2$, $E^p(\hat{G}) \subseteq E^2(\hat{G})$, so that $FT^{-1}\Psi$ is defined for every $\Psi \in E^p(\hat{G})$. The Hausdorff-Young inequality for $\hat{G}$ asserts that in fact $FT^{-1}\Psi$ belongs to $L^{p'}(G)$ (not merely to $L^2(G)$) and that

$$\|FT^{-1}\Psi\|_{p'} \le \|\Psi\|_p$$

for every $\Psi \in E^p(\hat{G})$; see Hewitt and Ross [1], (31.24).

There are also versions of the Hausdorff-Young inequality applying to any locally compact Abelian group; see loc. cit., (31.20).

2.14.7. Exercise. Assume that (2.14.1) has been established for every $p \in [1, 2]$ and every $f \in T(G)$. Write out a detailed proof of the fact that (2.14.1) continues to hold for every $p \in [1, 2]$ and every $f \in L^p(G)$.

[Hint: Use an approximate identity (k_j) whose elements belong to $T(G)$; cf. 2.9.6 above.]

2.14.8. Exercise. Let G be an infinite Abelian compact group with dual group Γ, Γ_0 an infinite subset of such that $\underline{1} = \underline{1}_G \in \Gamma_0$. Define $f_0 = g_0 = \underline{1}$. Show how to choose by recurrence elements $\chi_1, \chi_2, \ldots$ of Γ_0 such that, on writing

$$f_{n+1} = f_n + \chi_{n+1} g_n, \quad g_{n+1} = f_n - \chi_{n+1} g_n,$$

one obtains t.p.s f_n such that $\|f_n\|_\infty \le 2^{\frac{1}{2}n+\frac{1}{2}}$, Ran $\hat{f}_n \subseteq \{-1, 0, 1\}$, and $|\hat{f}_n|$ is the characteristic function of a subset Δ_n of the subsemigroup Γ_0^∞ of Γ generated by Γ_0, Δ_n having precisely 2^n elements.

Remark. The f_n are analogues of the Rudin-Shapiro trigonometric polynomials on the circle group T; see Katznelson [1], p. 33,

Exercise 6; Hewitt and Ross [1], (37.19.b). They have many interesting and useful properties, some of which are exhibited in the next two exercises.

2.14.9. Exercise. The notation is as in the preceding exercise. Show that one can choose by recurrence elements $\alpha_1, \alpha_2, \ldots$ of Γ_0 so as to arrange that, if

$$w_n = 2^{-\frac{1}{2}n} \alpha_n f_n ,$$

then $\operatorname{Ran} \hat{w}_n \subseteq \{-2^{-\frac{1}{2}n}, 0, 2^{-\frac{1}{2}n}\}$ and $2^{\frac{1}{2}n}|\hat{w}_n|$ is the characteristic function of a set $\Delta'_n \subseteq \Gamma_0^\infty$ having precisely 2^n elements, and $\Delta'_m \cap \Delta'_n = \phi$ whenever $m \neq n$.

Exhibit functions $g \in C(G)$ of the form

$$g = \Sigma_{n=0}^{\infty} c_n w_n , \tag{2.14.14}$$

where

$$\Sigma_{n=0}^{\infty} |c_n| < \infty , \tag{2.14.15}$$

such that

$$\Sigma_{\chi \in \Gamma} |\hat{g}(\chi)|^q = \infty$$

for every $q < 2$.

2.14.10. Exercise. The notations are as in the last two exercises. In addition, ϕ denotes any function on Γ such that $\operatorname{Ran}\phi \subseteq \{-1, 1\}$ and $\hat{w}_n$ agrees on Δ'_n with $2^{-\frac{1}{2}n}\phi$ for every n.

(i) Let (a_n) be any sequence of complex numbers such that

$$\limsup_{n\to\infty} 2^{\frac{1}{2}n} |a_n| = \infty .$$

Prove that there exists no function $f \in L^1(G)$ such that $\hat{f}$ agrees on Δ'_n with $a_n \phi$ for every n.

(ii) Let $\eta \geq 0_\Gamma$ belong to $c_0(\Gamma)$. Show how to construct complex-valued functions ψ on Γ such that

$$\Sigma_{\chi \in \Gamma} \eta(\chi) |\psi(\chi)|^2 < \infty$$

and yet ψ is not the Fourier transform of any integrable function on G. [Hints: For (i), consider functions $g * f$, where (2.14.14) and (2.14.15) hold, and make use of 2.8.2 and 2.8.4. For (ii), make judicious use of (i).]

Remark. It is not difficult to modify the arguments to cover the case in which $L^1(G)$ is replaced throughout by $M(G)$; cf. Remark (ii) following 2.8.4.

2.15. Lacunarity

2.15.0. With the possible exception of a few topics mentioned briefly in 2.13, we have so far dealt only with some of the very basic and relatively stabilised aspects of harmonic analysis on compact groups. There is no space to tackle, even in sketchiest outline, a representative collection of more specialised topics, in most of which there is as yet little sign of finality. We may, however, attempt to indicate the state of affairs in one such specialised topic, namely, lacunarity. This is one of the current foci of interest and may serve as an illustration.

We shall frequently confine our remarks to the Abelian case, for which a convenient reference is Chapter 15 of Edwards [3]. This case is certainly adequate to present many of the essential features. On the other hand, it is by no means always easy to extend results from the Abelian case to the non-Abelian one (indeed, such an extension is not always possible... see 2.15.4 below). All the same, when the going would become extra-difficult, or when complications look like mounting, we shall refer the reader to Hewitt and Ross [1], §37 for details of the non-Abelian case. It is in any case fair to say that, when it comes down to specific examples, available knowledge on the non-Abelian case is very sparse (almost lacunary, in fact). The Abelian case is already enough to handle!

2.15.1. Spectral subspaces. If $f \in L^1(G)$ (or $M(G)$), we shall here term spectrum of f, written $sp(f)$, the set of $U \in \hat{G}$ such that $\hat{f}(U) \neq 0$. (When G is Abelian, we shall always write Γ in place of $\hat{G}$;

see 2.5.4 above.) When G is non-Abelian, this concept of spectrum is not the same as any of the left- or right-E-spectra defined in 2.11.4 above; there is complete accord when G is Abelian, however, as is indicated in 2.11.5.

A function $f \in L^1(G)$ (or measure $\in M(G)$) is said to be S-spectral, where S is a subset of $\hat{G}$ (or of Γ), if and only if $sp(f) \subseteq S$. If E denotes (cf. 2.10.1) $C(G)$, $L^p(G)$ or $M(G)$, E_S will denote the set of $f \in E$ which are S-spectral. Plainly, E_S is a closed linear subspace of E (actually a closed ideal in E). These subspaces E_S may be referred to as spectral subspaces of E.

By a trigonometric series on G is meant a series

$$\Sigma_{U \in \hat{G}} d(U). \operatorname{Tr}[\Psi(U)U(x)^*],$$

where $\Psi \in \Pi_{U \in \hat{G}} \operatorname{End}(\mathscr{H}_U)$; if G is Abelian, this boils down to the form

$$\Sigma_{\chi \in \Gamma} c(\chi)\chi ;$$

where c is a complex-valued function on Γ. (In either case the series are at present purely formal: convergence or summability are notions to be defined with care.) If $S \subseteq \hat{G}$, the series is termed S-spectral if and only if $\Psi(U) = 0$ for every $U \in \hat{G} \backslash S$.

It turns out that, if S is suitably sparse... lacunary (in the sense of being broken by large gaps) appears to be the hallowed term... individual elements of E_S, and E_S itself, exhibit properties strikingly different from those of general elements of E, or of E itself; and S-spectral trigonometric series behave quite differently from trigonometric series in general. This is the phenomenon of lacunarity.

2.15.2. **Hadamard sets.** The earliest instance of lacunarity with some semblance of generality was probably that linked with the odd properties of power series in one complex variable which exhibited so-called Hadamard gaps, i.e., series

$$\Sigma_{k=1}^{\infty} c_k z^{n_k}$$

in which the positive integers n_k are such that

$$q = \inf_k n_{k+1}/n_k > 1 . \tag{2.15.1}$$

(Weierstrass' famous continuous nowhere differentiable function is, however, an even earlier instance of a trigonometric series exhibiting gaps of the same sort.) Since the peculiar behaviour of such power series centres around their behaviour on or near their circle of convergence (which may without loss of generality be taken to be the unit circle), it was entirely natural to pass from the power series to the trigonometric series which formally represents its boundary values, and thence to two-way infinite lacunary trigonometric series on the circle group **T**. Such a series, say

$$\Sigma_{n \in Z} c_n e^{int} , \tag{2.15.2}$$

is said to exhibit Hadamard gaps, or to be a Hadamard series, if $c_n = 0$ for every $n \in \mathbf{Z}$ save perhaps for those of the form $n = \pm n_k$ or 0, where the positive integers n_k satisfy (2.15.1). Likewise, a subset S of **Z** which is contained in a set of the form $\{0\} \cup \{\pm n_k : k=1, 2, \dots\}$, where the n_k are positive and satisfy (2.15.1), is usually termed a Hadamard set.

Quite a number of special properties of S-spectral functions and trigonometric series on **T**, valid whenever S is a Hadamard set of integers, were discovered (by Sidon, F. Riesz and others) before there was any attempt (mainly due to Banach) to try to characterise species of lacunary subsets of **Z** directly in terms of the odd behaviour of S-spectral functions or S-spectral subspaces (rather than in terms involving arithmetical operations and concepts applying to integers, as in the case of Hadamard sets). Once this was attempted, however, the ideas inevitably extended to general orthogonal expansions and to Fourier series on compact groups in particular. What is one of the best-known and strongest species of lacunarity specified in this way has come to be linked with the name of Sidon. We will describe this concept in the general group setting, bearing in mind that in this context **Z** is always to be identified with the dual of **T** (recall Exercise 2.2.15).

2.15.3. Sidon sets. A subset S of $\hat{G}$ is said to be a Sidon set if and only if either of the following two equivalent conditions is fulfilled:

(s_1) $L_S^\infty(G)$ (or $C_S(G)) \subseteq A(G)$, i.e., $\hat{f} \in E^1(\hat{G})$ whenever $f \in L_S^\infty(G)$ (or whenever $f \in C_S(G)$);

(s_2) there exists a number $\kappa = \kappa(S)$ such that

$$\|\hat{f}\|_1 \le \kappa \|f\|_\infty$$

for every $f \in T_S(G)$ (or $C_S(G)$, or $L_S^\infty(G)$).

It turns out that S is Sidon if and only if it satisfies either of the following further two conditions:

(s_3) if $\Psi \in E^\infty(G)$, there exists $\lambda \in M(G)$ such that

$$\Psi(U) = \hat{\lambda}(U) \text{ for every } U \in S;$$

(s_4) if $\Psi \in E_0(\hat{G})$, there exists $f \in L^1(G)$ such that $\Psi(U) = \hat{f}(U)$ for every $U \in S$.

For proofs of the equivalence of all these conditions, see Hewitt and Ross [1], (37.2); for the Abelian case, see also Edwards [3], 15.1.4.

In case $G = \mathbf{T}$, all Hadamard subsets of $\mathbf{Z}$ are Sidon sets (see, for example, Edwards [3], 15.2.4); the converse is false (loc. cit., Exercise 15.3).

Evidently, (s_1) and (s_2) express special properties of certain S-spectral functions and S-spectral subspaces; they assert that certain familiar norms assume unexpected properties, when they are restricted to S-spectral functions (or S-spectral subspaces). On the other hand, (s_3) and (s_4) are of a superficially quite different sort; they may be said to express 'covering' or 'matching' properties. (Compare them with the substance of 2.14.3.) The connections between these two apparently different sorts of properties were first displayed by Banach, whose arguments indicated a very general sort of 'principle of duality'; see the discussion on pp. 525-32 of Edwards [2].

Several other strange properties of Sidon sets may as well be stated here:

(s') there exists a number $\kappa_1 = \kappa_1(S)$ such that

$$\|f\|_p \leq \kappa_1 p^{\frac{1}{2}} \|f\|_1$$

for every $f \in L^1_S(G)$ and every $p \in [1, \infty)$;

see Hewitt and Ross [1], (37.25) and Edwards [3], 15.3.1 for the Abelian case. A consequence of (s') is that, if S is a Sidon set, then every S-spectral measure λ is of the form $\lambda = \mu^f$ for some f which belongs to $L^p(G)$ for every finite p. (The formula $\lambda = \mu^f$ means that $\lambda(g) = \int gf d\mu$ for every $g \in C(G)$; see 2.1.4(iii). This may be shown to signify that λ is absolutely continuous with respect to Haar measure μ and the Radon-Nikodym derivative of λ with respect to μ is f.) Further, taking G to be Abelian for simplicity,

(s") if $\psi \in l^2(S)$, there exists $f \in C(G)$ such that $\hat{f}(\chi) = \hat{\psi}(\chi)$ for every $\chi \in S$;

(s‴) there exists a number $\kappa_1 = \kappa_1(S)$ such that, if $f \in L^p(G)$ and $p > 1$, then

$$\{\Sigma_{\chi \in S} |\hat{f}(\chi)|^2\}^{\frac{1}{2}} \leq \kappa_1 p'^{\frac{1}{2}} \|f\|_p ;$$

for proofs, see Edwards [3], 15.3.2 and 15.3.3.

2.15.4. General comments

(i) As has been said, each of the statements (s'), (s") and (s‴) is true of every Sidon set S. Presumably, no one of them implies that S is Sidon, though the writer knows of no proof of this.

A set $S \subseteq \hat{G}$ such that (s') is true for a given $p \in (1, \infty)$, the number $\kappa_1 p^{\frac{1}{2}}$ being replaced by an unspecified number $\kappa' = \kappa'(S, p)$, is termed <u>a set of type</u> Λ_p, or a Λ_p<u>-set</u>; see Hewitt and Ross [1], (37.6) and Edwards [3], 15.5. (In the latter reference, Λ_p is defined for every $p \in (0, \infty)$.) There are matching properties (analogous to (s_3) and (s_4)) which serve to characterise Λ_p-sets; see the references just cited. These Λ_p-sets form another recognised species of lacunary sets. A set which is of type Λ_p is also of type Λ_q for every $q \in (0, p)$.

In view of (s'), every Sidon set is a Λ_p-set for every finite p. Examples are known (Edwards, Hewitt and Ross [1], [2]) of sets which are Λ_p-sets for every finite p and which are not Sidon sets. However, it is apparently unknown whether any such set satisfies (s'), inasmuch as the variation of $\kappa'(S, p)$ with p may not be in accord with that specified in (s').

(ii) If G is Abelian, every infinite subset of Γ contains an infinite Sidon set. (The proof for $G = \mathbf{T}$ in Edwards [3], 15. 2. 5 can be modified so as to apply to a general Abelian G; see Hewitt and Ross [1], (37. 18).) This assertion is, however, not true for a general non-Abelian compact G: in this case, indeed, it may happen that no infinite subset of $\hat{G}$ is of type Λ_4 (Hewitt and Ross [1], 37. 21. b)).

(iii) Using the positive part of (ii), one can show that, if G is Abelian, the statements (s') and (s''') cannot be too much strengthened without forcing S to be finite. More specifically, for no infinite set $S \subseteq \Gamma$ is it true either that

$$\bigcap_{p<\infty} L^p_S(G) \subseteq L^\infty(G)$$

or that

$$\hat{f}\,|\,S \in \bigcup_{q<\infty} l^q(S) \quad \text{for every } f \in L^1(G) ;$$

the first assertion is a special case of Edwards [3], Exercise 15. 15, and the second is dealt with in Exercise 2. 15. 5 below.

On the other hand, (s') is enough to show that, if S is a Sidon set, then

$$\int_G \exp(c\,|f|^2)d\mu < \infty$$

for every $f \in L^1_S(G)$ and every real number c (Edwards [3], Exercise 15. 4; cf. Exercise 2. 15. 6 below). An S-spectral L^1-function can therefore be but 'mildly unbounded', if S is Sidon.

(iv) The criteria (s_1)-(s_4) for Sidon sets (and the analogous criteria for Λ_p-sets), although satisfactory in some respects, are very difficult to apply in practice. If G is non-Abelian, examples of Sidon or Λ_p-sets are very hard to come by. Greater variety exists when G is Abelian, but in this case virtually every claim that a specified set is or is not Sidon (or of type Λ_p) is, if justifiable at all, justified by reference to certain arithmetical or group-theoretical properties; cf. Edwards [3],

15.2.2, 15.2.8, 15.5.5, Exercises 15.7-15.9 and Hewitt and Ross [1], (37.14) and (37.27).

Looking a little more closely at the Abelian case, it turns out that most properties of this type presently available are expressed in terms of a function $\alpha_S : \mathbf{N} \mapsto \mathbf{N}$ defined by

$$\alpha_S(n) = \sup \#(S \cap \{\chi \xi^k : k = 0, 1, 2, \ldots, n-1\}),$$

where, for any finite set F, $\#F$ denotes the cardinal of F, and where the supremum is taken with respect to all $\chi \in \Gamma$ and to all $\xi \in \Gamma$, $\xi \neq \underline{1}_G$. If $G = \mathbf{T}$, $\alpha_S(n)$ is the largest integer α such that some arithmetic progression of n terms contains α elements of S. Sticking to this special choice of G (though a good deal of what is to be said can be extended to cover cases in which G is compact Abelian), one has (see Edwards [3], loc. cit.) the following facts:

$$\alpha_S(n) = \underline{0}(\log n) \text{ if } S \text{ is a Sidon set,}$$

and one can have $\alpha_S(n) > c.\log n$ for some $c > 0$, every n and suitable Sidon sets S;

$$\alpha_S(n) = \underline{0}(n^{p/2}) \text{ if } S \text{ is a } \Lambda_p\text{-set and } p > 2;$$

$$\alpha_S(n) < n \text{ for all sufficiently large } n \text{ if } S \text{ is of type } \Lambda_1.$$

Moreover, if $\alpha : \mathbf{N} \mapsto \mathbf{N}$ is such that $\alpha(n) = \underline{0}(n^\varepsilon)$ for every $\varepsilon > 0$, one can construct sets $S \subseteq \mathbf{Z}$ which are of type Λ_p for every $p \in (0, \infty)$ and for which $\alpha_S(n) > \alpha(n)$ for an infinite of $n \in \mathbf{N}$. (The writer does not know whether this last property is valid when G is any compact Abelian group.) In this way one can obtain many sets S which are of type Λ_p for every finite p and which are not Sidon sets.

Specialising yet more, consider the case in which $G = \mathbf{T}$ and S has the form $\{0\} \cup \{\pm n_k : k = 1, 2, \ldots\}$, where $0 < n_1 < n_2 < \ldots$. The results cited in the last paragraph show, for example, that $n_k - n_h \geq \exp(c(k-h))$ for some $c = c(S) > 0$ and every $k > h$, whenever S is Sidon. In particular, the differences $n_{k+1} - n_k$ must be 'large on the average' (though they may be no larger than 1 for an infinity of k). A similar inference is valid when S is of type Λ_p and $p > 2$.

There is some evidence that partition-type (or representation-type) problems familiar in additive number theory would have some bearing upon lacunarity; yet, as far as the writer is aware, nothing very specific has been achieved in this direction (cf. Edwards [3], 15. 5. 5).

Thus, notwithstanding the very useful facts recited above, there remains a great need for lots of well-chosen examples of Sidon sets and of Λ_p-sets. (In this connection, see the relevant sections of Meyer [1].) The need is especially acute in case G is non-Abelian.

(v) The situation regarding convergence properties of S-spectral trigonometric series is also unsatisfactory. On the one hand, there are numerous results known which state strange convergence properties of Hadamard series on the circle group (see, for example, Edwards [3], Exercise 15. 17 and the references cited there); and also some extensions of these to various species of lacunary trigonometric series on general compact Abelian groups (though the writer thinks that few of these have ever been published). On the other hand, there seem to be no really satisfactory results of the converse type, i. e., ones which assert that, if every S-spectral series behaves in this or that strange fashion in relation to convergence, then S must be lacunary in some preassigned sense.

(vi) That there remain large gaps in the existing knowledge about various species of lacunarity, is already quite plain. It is easy to make this even plainer by means of one or two further questions which are still unanswered.

Taking a fixed G, how does the set of Sidon sets (or the set of Λ_p-sets for given p) behave in relation to finite unions? Drury [1] has recently shown that the union of two Sidon sets is again a Sidon set; see also Edwards [3], Exercise 15. 10 and Hewitt and Ross [1], (37. 21. a). It is evident that the set of Λ_p-sets does not become enlarged when p increases; does it decrease strictly when p increases?

Assuming G to be Abelian, how does the set of Sidon sets (or of Λ_p-sets for given p) behave in relation to the group-theoretical product of subsets of $\mathbf{\Gamma}$? If $G = \mathbf{T}$, it is known that on the one hand the sum of two Hadamard sets of non-negative integers is of type Λ_p for every $p < \infty$ (Meyer [1], p. 558); and on the other hand that the sum of two infinite sets is never a Sidon set (Kahane [1], p. 61, Exercise 4).

Any interested reader will be able to extend the list.

2.15.5. **Exercise.** Suppose that G is compact Abelian and that S is a subset of Γ such that to every $f \in L^1(G)$ corresponds a real number $q = q(f) \geq 1$ for which

$$\Sigma_{\chi \in S} |\hat{f}(\chi)| < \infty .$$

Prove that S is finite.

[Hints: For $q = 1, 2, \ldots$ define $P_q : L^1(G) \to [0, \infty]$ by $P_q(f) = (\Sigma_{\chi \in S} |\hat{f}(\chi)|^q)^{1/q}$. Apply the uniform boundedness principle (Edwards [3], Appendix B.2.1) to the P_q to conclude that there exists a positive integer q and a number $m \in (0, \infty)$ such that $P_q(f) \leq m\|f\|_1$ for every $f \in L^1(G)$. Using Appendix C.1 loc. cit., conclude that to every $\psi \in l^{q'}(\Gamma)$ which vanishes on $\Gamma \backslash S$ corresponds $g \in L^\infty(G)$ such that $\hat{g} = \psi$. Now use 2.15.3(s_1) and 2.15.4(ii).]

Note. The final exercise to follow provides the interested reader with a lead-in to the study of so-called random series. It involves an analogue of Sections 14.2 and 14.3 of Edwards [3], the differences being that the role played there by the Rademacher functions is here taken over by a lacunary set of characters of a suitable compact Abelian group K, and the essential features of 14.2.1 and 14.2.2 being replaced by 2.15.3(s') above.

For a deep and general study of random series, see Kahane [1]. Some aspects of the theory for general compact groups, due largely to Helgason, Figà-Talamanca and Rider, appear in §36 of Hewitt and Ross [1].

2.15.6. **Exercise.** Let $G = \{x\}$ and $K = \{t\}$ be infinite compact Abelian groups with duals $\Gamma = \{\chi\}$ and $\Omega = \{\omega\}$ respectively. Take $c \in l^2(\Gamma)$ and choose elements χ_n $(n = 1, 2, \ldots)$ of Γ such that $c(\chi) = 0$ for every $\chi \in \Gamma$ different from every χ_n; write c_n for $c(\chi_n)$. Let $n \mapsto \omega_n$ be an injection of $\{1, 2, \ldots\}$ into Ω such that $S = \{\omega_n : n = 1, 2, \ldots\}$ is a Sidon subset of Ω (see 2.15.4(ii)). Write

$$s_n(x, t) = \Sigma_{r=1}^{n} c_r \omega_r(t) \chi_r(x) .$$

For each $t \in K$, let f_t denote an element of $\mathscr{L}^2(G)$ which is the limit, in $\mathscr{L}^2(G)$ and as $n \to \infty$, of the sequence of functions $x \mapsto s_n(x, t)$. Starting from 2.15.3(s'), applied with K in place of G, and following the argument in Section 14.3 of Edwards [3], show that there is a μ_K-negligible set $E \subseteq T$ such that, for every $t \in K\backslash E$,

$$\int \exp(k|f_t|^2)d\mu_G < \infty \qquad (2.15.3)$$

for every $k \in (0, \infty)$; a fortiori, $f_t \in \mathscr{L}^p(G)$ for every $t \in K\backslash E$ and every finite p.

Remarks. (i) By choosing K to be a denumerably infinite power of the two-element multiplicative group $\{-1, 1\}$, it can be arranged that Ran $\omega \subseteq \{-1, 1\}$ for every $\omega \in \Omega$; cf. Remark 2.2.13 above.

(ii) Starting from an $f \in \mathscr{L}^2(G)$ such that $f \notin \bigcup_{s>2} \mathscr{L}^s(G)$ (see Exercise 2.1.16), and taking $c = \hat{f}$, one obtains a function g $(= f_t$ for any chosen $t \in K\backslash E)$ such that $g \in \mathscr{L}^p(G)$ for every finite p and yet, for some choice of the $\pm$ signs, the trigonometric series

$$\Sigma_{\chi \in \Gamma} \pm g(\chi)\chi$$

fails to be the Fourier series of any function belonging to $\bigcup_{s>2} \mathscr{L}^s(G)$. (Other results of a similar type are to be found in Edwards [3], Chapter 14 and in the references cited there.) This has bearing upon another central specialised topic in harmonic analysis, namely, the study of multipliers; see loc. cit. Chapter 16.

(iii) Assume merely that $c \in l^2(\Gamma)$ is given. Then, although (2.15.3) is proven for every $t \in K\backslash E$ for some $E \subseteq K$ satisfying $\mu_K(E) = 0$ (and hence surely for uncountably many $t \in K$), it is both salutary and curious that (as far as the writer is aware) E remains completely unspecified in all other respects. As a consequence, there is no known procedure applicable when c is an arbitrarily given element of $l^2(\Gamma)$ which will lead to a single specific $t \in K$ for which (2.15.3) may be guaranteed.

Concluding remarks. Readers who wish to pursue harmonic analysis may find Edwards [3] a useful guide to topics not mentioned in these notes and to some books and research papers. To the books about harmonic analysis listed in the bibliography of Edwards [3] should be added Hewitt and Ross [1], Volume II (used throughout these notes time and again), Bourbaki [2], [3], Reiter [1], Katznelson [1], Kahane [1] and [2], Benedetto [1], Donoghue [1], Ehrenpreis [1]. As for research papers, we can only suggest to the reader that he makes frequent use of reviewing journals!

For historical remarks, see the notes at the ends of sections in Hewitt and Ross [1], and the historical note attached to Bourbaki [2].

Appendix A

Notation. $\mathscr{H}$: a Hilbert space of f. d. d, scalar product $(\ |\ )$, norm $\|.\|$;

$\underset{\sim}{a}, \underset{\sim}{b}, \ldots, \underset{\sim}{u}, \underset{\sim}{v}$: vectors (elements of $\mathscr{H}$);

$\mathscr{E} = \text{End}\,(\mathscr{H})$, I the identity endomorphism of $\mathscr{H}$,

A, B, ..., X: elements of $\mathscr{E}$;

$$\|A\| = \sup\{\, \|A\underset{\sim}{a}\| : \|\underset{\sim}{a}\| \le 1 \,\} ;$$

$\alpha, \beta, \ldots, \lambda$: scalars.

A. 0. Adjoints

A. 0. 1. If $A \in \mathscr{E}$ its (Hilbert) adjoint is the element A^* of $\mathscr{E}$ defined by

$$(A\underset{\sim}{a}|\underset{\sim}{b}) = (\underset{\sim}{a}|A^*\underset{\sim}{b}) \qquad (A.\ 0.\ 1)$$

for every $\underset{\sim}{a}, \underset{\sim}{b} \in \mathscr{H}$.

$A \mapsto A^*$ is a conjugate-linear isometry of $\mathscr{E}$ onto itself such that $A^{**} = (A^*)^* = A$ and $(AB)^* = B^*A^*$.

A. 0. 2. A is said to be self-adjoint (s. a. for short) if and only if $A^* = A$ (the terms Hermitian and symmetric are sometimes used in place of self-adjoint). When A is s. a., $(A\underset{\sim}{a}|\underset{\sim}{a})$ is real for every $\underset{\sim}{a} \in \mathscr{H}$; if further $(A\underset{\sim}{a}|\underset{\sim}{a}) \ge 0$ for every $\underset{\sim}{a} \in \mathscr{H}$, A is said to be positive self-adjoint (p. s. a.). This concept of positivity leads to a partial order on the set of s. a. endomorphisms.

For every $A \in \mathscr{E}$, AA^* and A^*A are p. s. a; and $A = A_1 + iA_2$, where $A_1 = \frac{1}{2}(A + A^*)$ and $A_2 = \frac{1}{2}i(A^* - A)$ are s. a.

A. 0. 3. An element U of $\mathscr{E}$ is termed unitary if and only if it is invertible in $\mathscr{E}$ and $U^{-1} = U^*$. U ($\mathscr{H}$) denotes the set of unitary $U \in \mathscr{E}$.

A. 0. 4. An element A of $\mathscr{E}$ is termed normal if and only if $AA^* = A^*A$.

A. 0. 5. An (orthogonal) projector on (or in) $\mathscr{H}$ is an element P of $\mathscr{E}$ such that P is s. a. and indempotent ($P^2 = P$). For any $\underset{\sim}{a} \in \mathscr{H}$, $P\underset{\sim}{a}$ is then the (orthogonal) projection of $\underset{\sim}{a}$ onto the subspace $P(\mathscr{H}) = \text{Ran } P$, i. e., $P\underset{\sim}{a}$ is the unique $\underset{\sim}{u} \in \mathscr{H}$ such that $\underset{\sim}{a} - \underset{\sim}{u}$ is orthogonal to Ran P. The dimension of P is that of Ran P. Projectors P and P' are said to be orthogonal if and only if Ran P and Ran P' are orthogonal subspaces of $\mathscr{H}$, which is so if and only if $PP' = 0$ (in which case $P'P = 0$ as well).

A. 0. 6. If A is s. a.,

$$\|A\| = \sup \{ |(A\underset{\sim}{a}|\underset{\sim}{a})| : \|\underset{\sim}{a}\| \le 1 \}$$

Proof. Denote the sup on the right by m. Plainly $m \le \|A\|$. To prove the reverse inequality, note that, since A is s. a.,

$$\text{Re } (A\underset{\sim}{a}|\underset{\sim}{b}) = (A\underset{\sim}{c}|\underset{\sim}{c}) - (A\underset{\sim}{d}|\underset{\sim}{d}) ,$$

where $\underset{\sim}{c} = \frac{1}{2}(\underset{\sim}{a}+\underset{\sim}{b})$, $\underset{\sim}{d} = \frac{1}{2}(\underset{\sim}{a}-\underset{\sim}{b})$. Hence

$$\begin{aligned} |\text{Re } (A\underset{\sim}{a}|\underset{\sim}{b})| &\le m(\|\underset{\sim}{c}\|^2 + \|\underset{\sim}{d}\|^2) \\ &= \tfrac{1}{2}m(\|\underset{\sim}{a}\|^2 + \|\underset{\sim}{b}\|^2) . \end{aligned}$$

Take non-zero vectors $\underset{\sim}{u}$ and $\underset{\sim}{v}$ and replace $\underset{\sim}{a}$ and $\underset{\sim}{b}$ in the above by $\|\underset{\sim}{v}\| \, \|\underset{\sim}{u}\|^{-1}\theta\underset{\sim}{u}$ and $\|\underset{\sim}{u}\| \, \|\underset{\sim}{v}\|^{-1}\underset{\sim}{v}$ respectively, where $|\theta| = 1$ is chosen so that $\theta(A\underset{\sim}{u}|\underset{\sim}{v}) = |(A\underset{\sim}{u}|\underset{\sim}{v})|$. It then appears that

$$|(A\underset{\sim}{u}|\underset{\sim}{v})| \le \tfrac{1}{2}m(\|\underset{\sim}{v}\|^2 + \|\underset{\sim}{u}\|^2) .$$

The same is evidently true if either or both of $\underset{\sim}{u}$, $\underset{\sim}{v}$ is zero. Hence $|(A\underset{\sim}{u}|\underset{\sim}{v})| \le m$ whenever $\underset{\sim}{u}, \underset{\sim}{v} \in \mathscr{H}$ have norms not greater than unity.

Homogeneity then shows that $|(\underset{\sim}{A u} | \underset{\sim}{v})| \leq m \|\underset{\sim}{u}\| \, \|\underset{\sim}{v}\|$ whenever $\underset{\sim}{u}, \underset{\sim}{v} \in \mathscr{H}$. Taking $\underset{\sim}{v} = A\underset{\sim}{u}$, it follows that $\|A\underset{\sim}{u}\|^2 \leq m \|\underset{\sim}{u}\| \, \|A\underset{\sim}{u}\|$, and so $\|A\underset{\sim}{u}\| \leq m \|\underset{\sim}{u}\|$ for every $\underset{\sim}{u} \in \mathscr{H}$, showing that $\|A\| \leq m$.

A. 0. 7. If $A \in \mathscr{E}$ and λ is a scalar, write

$$M(A, \lambda) = \{\underset{\sim}{a} \in \mathscr{H} : A\underset{\sim}{a} = \lambda \underset{\sim}{a}\} .$$

Then $M(A, \lambda)$ is a linear subspace of $\mathscr{H}$ termed the <u>eigenmanifold</u> or <u>spectral manifold</u> of A associated with λ. λ is said to be an <u>eigenvalue</u> of A if and only if $M(A, \lambda) \neq \{\underset{\sim}{0}\}$; and then every $\underset{\sim}{a} \in M(A, \lambda) \setminus \{\underset{\sim}{0}\}$ is termed an <u>eigenvector</u> of A associated with λ. The set

$$\sigma(A) = \{\lambda \in \mathbf{C} : M(A, \lambda) \neq \{\underset{\sim}{0}\}\}$$

of eigenvalues of A is termed the <u>spectrum of</u> A.

If A is s. a. (resp. p. s. a.), its spectrum is real (resp. real and non-negative) and $M(A, \lambda)$ and $M(A, \lambda')$ are orthogonal whenever $\lambda \neq \lambda'$.

If U is unitary, every element of $\sigma(U)$ is unimodular.

A. 0. 8. All the preceding definitions and results carry over when $\mathscr{H}$ is infinite dimensional, provided 'endomorphism of $\mathscr{H}$' is replaced throughout by 'continuous endomorphism of $\mathscr{H}$', and $\mathscr{E}$ is the set of such continuous endomorphisms of $\mathscr{H}$.

A. 1. The spectral theorem

A. 1. 1. By a <u>spectral family</u> is meant a family $\lambda \in \mathbf{C} \mapsto P_\lambda$ of projectors on $\mathscr{H}$ such that

(i) $P_\alpha P_\beta = 0$ if $\alpha \neq \beta$;

(ii) $\Sigma_{\lambda \in \mathbf{C}} P_\lambda = I$.

(It is a consequence of (i) that $P_\lambda \neq 0$ for but a finite set of λ, so that the sum in (ii) is a finite one.)

If $A \in \mathscr{E}$, such a spectral family is said to <u>belong to</u> A, or to form a <u>spectral resolution of</u> A, if and only if

$$A = \Sigma_{\lambda \in \mathbf{C}} \lambda P_\lambda . \tag{A. 1. 1}$$

In view of (i), (A. 1. 1) can hold only if A is normal. Furthermore, (A. 1. 1) determines the spectral family (P_λ) uniquely (whenever it exists): P_λ must be the orthogonal projector of $\mathscr{H}$ onto the eigen-manifold $M(A, \lambda)$.

A. 1. 2. Spectral theorem. If $A \in \mathscr{E}$ is normal, there exists a (necessarily unique) spectral resolution (P_λ) of A.

For a proof, see Halmos [2], Sections 79, 80.

Note that the sum in (iii) can be restricted to $\sigma(A)$ or to $\sigma(A) \setminus \{0\}$ (since $\lambda P_\lambda = 0$ for every other $\lambda \in \mathbf{C}$).

A. 1. 3. If A is normal and (P_λ) its spectral resolution, and if f is any complex-valued function defined on $\sigma(A)$, $f(A) \in \mathscr{E}$ is defined by

$$f(A) = \Sigma_{\lambda \in \sigma(A)} f(\lambda) P_\lambda ; \qquad \text{(A. 1. 2)}$$

properties (i)-(ii) in A. 1. 1 and (A. 1. 1) guarantee that this is a sensible definition (inasmuch as it gives the right answer when f is a polynomial function with complex coefficients). Since $\sigma(A)$ is finite, it is easy to see that $f(A)$ is always a polynomial in A with complex coefficients (even though f may not be a polynomial function). In particular, $f(A)$ is normal and commutes with A.

A. 1. 4. An important special case of A. 1. 3 is the definition of $A^{\frac{1}{2}}$ for every p. s. a. A: here $\sigma(A) \subseteq [0, \infty)$ and one takes $f(\lambda) = \lambda^{\frac{1}{2}}$ for λ real and non-negative, where $\lambda^{\frac{1}{2}}$ is again real and non-negative. Then $A^{\frac{1}{2}}$ is p. s. a. and its square is A; these properties characterise $A^{\frac{1}{2}}$ (see Halmos [2], p. 166).

A. 1. 5. For any $A \in \mathscr{E}$, AA^* is p. s. a. and $|A|$ is understood to be $(AA^*)^{\frac{1}{2}}$.

A. 1. 6. If $A \in \mathscr{E}$, it is possible to write $A = |A| U$, where $U \in U(\mathscr{H})$. For a proof, see Halmos [2], Section 83, remark following the proof of Theorem 1.

A. 2. The trace function

A. 2. 1. There exists precisely one complex-valued linear function Tr on $\mathscr{E}$ with the following properties:

(i) Tr is unitarily invariant, i. e., $\mathrm{Tr}(U^{-1}AU) = \mathrm{Tr}\, A$ for $U \in U(\mathscr{H})$ and $A \in \mathscr{E}$;

(ii) $\mathrm{Tr}\, I = d$.

Proof. As for uniqueness, (i) and (ii) show that $\mathrm{Tr}\, P = 1$ for every one-dimensional projector P. On the other hand, by A. 3. 3 below, every $A \in \mathscr{E}$ is a linear combination of such projectors.

As for existence, there are various ways of exhibiting a function Tr with the desired properties. One possibility appears as (2. 4. 4') of the main text (i. e., (A. 2. 1) below), though this is not very satisfying because of its use of an arbitrary choice of ONB in $\mathscr{H}$. Another way, singularly appropriate for these notes, rests on the observation that $U(\mathscr{H})$ is a compact group: let m denote normalised invariant measure on $U(\mathscr{H})$ and note that invariance of m ensures that

$$\int_{U(\mathscr{H})} (UAU^{-1}\underset{\sim}{e} \,|\, \underset{\sim}{e})\, dm(U)$$

is the same for every choice of $\underset{\sim}{e} \in \mathscr{H}$ satisfying $\|\underset{\sim}{e}\| = 1$. This same invariance of m ensures that

(iii) $\mathrm{Tr} : A \mapsto d\int_{U(\mathscr{H})} (UAU^{-1}\underset{\sim}{e} \,|\, \underset{\sim}{e})\, dm(U)$

has all the desired properties.

Whatever definition is used, the validity of the formula

$$\mathrm{Tr}\, A = \sum_{i=1}^{d} (A\underset{\sim}{e}_i \,|\, \underset{\sim}{e}_i) \qquad \text{(A. 2. 1)}$$

for every $A \in \mathscr{E}$ and every ONB $(\underset{\sim}{e}_i)_{i=1}^{d}$ in $\mathscr{H}$ is a consequence of uniqueness. (It can also be derived from (iii) above by using the orthogonality relations for the group $U(\mathscr{H})$.)

Remark. Cf. Hewitt and Ross [1], (D. 16).

A. 2. 2. From A. 2. 1(iii) and reflection-invariance of m (or from (A. 2. 1)) it follows that $\mathrm{Tr}\, A^* = (\mathrm{Tr}\, A)^-$ (cf. (2. 2. 7)).

Since Tr P = dim P ($\mathscr{H}$) for every projector P, the spectral theorem A. 1. 2 shows that

$$\text{Tr A} = \Sigma\lambda.\ \dim P_\lambda(\mathscr{H}) \qquad \text{(A. 2. 1')}$$

for every normal A, (P_λ) being the spectral resolution of A. Thence (or otherwise) it follows that Tr A is real whenever A is s. a., and that Tr A $\geq$ 0 whenever A is p. s. a. (with strict inequality unless A = 0).

From (A. 2. 1),

$$\text{Tr AB} = \text{Tr BA} \qquad \text{(A. 2. 2)}$$

and hence

$$\text{Tr } A^{-1}BA = \text{Tr B}\ ; \qquad \text{(A. 2. 3)}$$

cf. (2. 2. 5) and (2. 2. 6).

Applying (A. 2. 1) with AB in place of A and writing $B\underset{\sim}{e}_i = \Sigma_j (B\underset{\sim}{e}_i | \underset{\sim}{e}_j)\underset{\sim}{e}_j$, it follows that

$$\text{Tr AB} = \Sigma_{i,j} (A\underset{\sim}{e}_i | \underset{\sim}{e}_j)(B\underset{\sim}{e}_j | \underset{\sim}{e}_i)\ . \qquad \text{(A. 2. 4)}$$

Changing B into B* and using (A. 0. 1):

$$\text{Tr AB*} = \Sigma_{i,j} (A\underset{\sim}{e}_i | \underset{\sim}{e}_j)\overline{(B\underset{\sim}{e}_i | \underset{\sim}{e}_j)}\ . \qquad \text{(A. 2. 5)}$$

An application of the Cauchy-Schwarz inequality leads from (A. 2. 5) to

$$|\text{Tr AB*}| \leq (\text{Tr AA*})^{\frac{1}{2}}(\text{Tr BB*})^{\frac{1}{2}}\ . \qquad \text{(A. 2. 6)}$$

In particular,

$$|\text{Tr AU}| \leq d^{\frac{1}{2}}(\text{Tr AA*})^{\frac{1}{2}} \qquad \text{(A. 2. 7)}$$

if $U \in U(\mathscr{H})$.

By (A. 2. 1) and (A. 2. 5) and the Cauchy-Schwarz inequality,

$$\|A\| \leq (\text{Tr AA*})^{\frac{1}{2}} \leq d^{\frac{1}{2}}\|A\|\ . \qquad \text{(A. 2. 8)}$$

Next,

$$\left.\begin{aligned}\operatorname{Tr} ABB^*A^* &\leq \|B\|^2 \operatorname{Tr} AA^* ,\\ \operatorname{Tr} ABB^*A^* &\leq \|A\|^2 \operatorname{Tr} BB^* ,\\ \operatorname{Tr} ABB^*A^* &\leq \operatorname{Tr} AA^*. \operatorname{Tr} BB^* .\end{aligned}\right\} \qquad \text{(A. 2. 9)}$$

For $L = \|B\|^2 I - BB^*$ is p. s. a.; hence so also is ALA^*; hence $\operatorname{Tr} ALA^* \geq 0$, which leads at once to the first inequality. Also, $\operatorname{Tr} BAA^*B^* = \operatorname{Tr} BA(BA)^* = \operatorname{Tr} (BA)^*BA = \operatorname{Tr} A^*B^*BA \leq \|B^*\|^2 \operatorname{Tr} A^*A$ (by what has just been proved), and the second inequality follows by (A. 2. 2), the formula $\|B^*\| = \|B\|$ and an interchange of A and B. The third ensues by appeal to (A. 2. 8).

It can be shown (see Hewitt and Ross [1], (D. 32)) that

$$\operatorname{Tr} |A| = \sup \{ |\operatorname{Tr} AX| : X \in \mathscr{E} , \|X\| \leq 1 \}. \qquad \text{(A. 2. 10)}$$

In particular,

$$|\operatorname{Tr} AU| \leq \operatorname{Tr} |A| \qquad \text{(A. 2. 11)}$$

when $U \in U(\mathscr{H})$. It is a consequence of (A. 2. 10) that

$$\operatorname{Tr}(AA^*)^{\frac{1}{2}} = \operatorname{Tr}(A^*A)^{\frac{1}{2}} ; \qquad \text{(A. 2. 12)}$$

recall that $|A|$ was defined in A. 1. 4 to be $(AA^*)^{\frac{1}{2}}$, so that $(A^*A)^{\frac{1}{2}} = |A^*|$.

A. 3. Some lemmas

A. 3. 1. Lemma. (i) Let $A \in \mathscr{E}$ be normal and suppose that $\operatorname{Tr} Af(A) \geq 0$ (resp. $= 0$) for every polynomial function f such that $f(A)$ is p. s. a. Then A is p. s. a. (resp. $=0$).

(ii) If $A \in \mathscr{E}$ and $\operatorname{Tr}(T^*TA) \geq 0$ for every $T \in \mathscr{E}$, then A is p. s. a.

Proof. (i) In terms of the spectral decomposition (A. 1. 1),

$$Af(A) = \Sigma \lambda f(\lambda). P_\lambda$$

and so

$$\mathrm{Tr}\, Af(A) = \Sigma \lambda f(\lambda).\dim P_\lambda(\mathscr{H});\qquad (A.3.1)$$

the sum may be taken over those eigenvalues $\lambda \neq 0$ of A. Then $\dim P_\lambda(\mathscr{H}) > 0$ for every λ. The requirement that f(A) be p. s. a. is equivalent to $f(\lambda) \geq 0$ for all $\lambda \neq 0$ in the spectrum of A. If (A. 3. 1) is ≥ 0 (resp. 0) for all such f, it is clear that $\lambda \geq 0$ (resp. 0) for all λ in the spectrum of A, so that A is p. s. a. (resp. 0).

(ii) The given condition is equivalent to $\mathrm{Tr}(S^*AS) \geq 0$ for every $s \in \mathscr{E}$. Using (A. 2. 1) with S and the e_i suitably chosen, this entails that $(Aa|a) \geq 0$ for every $a \in \mathscr{H}$. Replacing a by $a + \lambda b$ and letting λ vary, one concludes that $(Aa|b) = (Ab|a)^- = (A^*a|b)$ for every $a, b \in \mathscr{H}$. Hence $A = A^*$ and A is seen to be p. s. a.

A. 3. 2. Lemma. Let $A \in \mathscr{E}$ be normal. Then A is a finite linear combination of one-dimensional projectors P, each of which is of the form $P = c.PA = c.AP$ (c a non-zero scalar); these projectors may be chosen to be orthogonal in pairs.

Proof. Use the spectral resolution (A. 1. 1) of A, noting that the sum may be confined to those $\lambda \neq 0$ in the spectrum of A, and that

$$P_\lambda A = AP_\lambda = \lambda P_\lambda .$$

Each P_λ is a sum of finitely many 1-dimensional projectors P^i_λ (corresponding to a choice of orthonormal base for $P_\lambda(\mathscr{H})$) and then the P^i_λ are mutually orthogonal and $P^i_\lambda P_\lambda = P_\lambda P^i_\lambda = P^i_\lambda$. Hence

$$P^i_\lambda A = P^i_\lambda P_\lambda A = P^i_\lambda . \lambda P_\lambda = \lambda P^i_\lambda ,$$

and likewise $\lambda P^i_\lambda = AP^i_\lambda$. Since $\lambda \neq 0$, each P^i_λ has the stated properties.

There is a weakened form of A. 3. 2 which applies to arbitrary endomorphisms, namely

A. 3. 3. Lemma. Every endomorphism is a finite linear combination of projectors.

Proof. This is true for any normal endomorphism (Lemma A. 3. 2)

hence in particular of any s. a. endomorphism. On the other hand, every $A \in \mathscr{E}$ is a linear combination of s. a. endomorphisms (see A. 0. 2).

A. 3. 4. Lemma. Let $A \in \mathscr{E}$. The set of right (resp. left) multiples of A in $\mathscr{E}$ is closed in $\mathscr{E}$.

Proof. Finite-dimensional (linear) subspaces are closed.

A. 3. 5. Lemma. If $T \in \mathscr{E}$, $T \neq 0$, every endomorphism of $\mathscr{H}$ can be expressed in the form $\sum_{j=1}^{d} A_j T B_j$ with $A_j, B_j \in \mathscr{E}$; in other words, the only two-sided ideals in $\mathscr{E}$ are $\{0\}$ and $\mathscr{E}$.

Proof. We can choose an orthonormal base $(\underset{\sim}{e}_i)$ for $\mathscr{H}$ such that $\underset{\sim}{a} = T\underset{\sim}{e}_1 \neq \underset{\sim}{0}$. Then choose $A_j \in \mathscr{E}$ such that $A_j\underset{\sim}{a} = \underset{\sim}{e}_j$ for $1 \leq j \leq d$. Finally choose $B_j \in \mathscr{E}$ so that

$$B_j\underset{\sim}{e}_i = \delta_{ij}\underset{\sim}{e}_1 \qquad (1 \leq i, j \leq d) .$$

Then

$$(\Sigma_j A_j T B_j)\underset{\sim}{e}_i = (\Sigma_j A_j T)\delta_{ij}\underset{\sim}{e}_1 = A_i T\underset{\sim}{e}_1 = A_i\underset{\sim}{a} = \underset{\sim}{e}_i$$

for $1 \leq i \leq d$. Accordingly $\Sigma_j A_j T B_j = I$, and the result follows. (As is seen, one can even assume that the A_j - or the B_j - shall be fixed in advance in a fashion depending only on T, not on the endomorphism to be represented.)

A. 3. 6. Lemma. Let P be a 1-dimensional projector on $\mathscr{H}$, and let $A \in \mathscr{E}$ be such that $PA \neq 0$ (resp. $AP \neq 0$). For every $B \in \mathscr{E}$ there exists $T \in \mathscr{E}$ such that $PB = PAT$ (resp. $BP = TAP$).

Proof. The second part follows from the first by taking adjoints ($AP \neq 0$ signifies that $PA^* \neq 0$), so we deal with the former only. Now P has the form $P\underset{\sim}{x} = (\underset{\sim}{x} | \underset{\sim}{a})\underset{\sim}{a}$ where $\|\underset{\sim}{a}\| = 1$, and $PA \neq 0$ shows that $A^*\underset{\sim}{a} \neq 0$. Accordingly, T^* can be chosen to satisfy $T^*A^*\underset{\sim}{a} = B^*\underset{\sim}{a}$. Then $PAT\underset{\sim}{x} = (AT\underset{\sim}{x} | \underset{\sim}{a})\underset{\sim}{a} = (\underset{\sim}{x} | T^*A^*\underset{\sim}{a})\underset{\sim}{a} = (\underset{\sim}{x} | B^*\underset{\sim}{a})\underset{\sim}{a} = (B\underset{\sim}{x} | \underset{\sim}{a})\underset{\sim}{a} = PB\underset{\sim}{x}$ for all $\underset{\sim}{x}$, i. e. $PB = PAT$.

A. 3. 7. Lemma. If $A \in \mathscr{E}$, then $A = \Sigma_i AP_i$, where the P_i are mutually orthogonal one-dimensional projectors of the form $P_i = X_i A$ for some $X_i \in \mathscr{E}$.

Proof. Let $Z = \text{Ker } A$ and take an ONB $(\underset{\sim}{e}_i)$ for $\mathscr{H}$ such that $(\underset{\sim}{e}_i)_{i \leq s}$ is a base for Z and $(\underset{\sim}{e}_i)_{i > s}$ a base for the orthogonal complement $Z^{\perp}$ of Z. Let P_i be the projection of $\mathscr{H}$ onto the subspace generated by $\underset{\sim}{e}_i$. If $i > s$, $Z \subseteq \text{Ker } P_i$ and so $P_i = X_i A$ for some $X_i \in \mathscr{E}$. (For example, define X_i as follows: if $\underset{\sim}{u} \in \text{Ran } A$, $\underset{\sim}{u} = A\underset{\sim}{v}$ for some $\underset{\sim}{v}$, and $P_i \underset{\sim}{v}$ depends only on $\underset{\sim}{u}$; put $X_i \underset{\sim}{u} = P_i \underset{\sim}{v}$; define X_i on the orthogonal complement $(\text{Ran } A)^{\perp}$ in any way so as to be linear.)

For any $\underset{\sim}{u}$,

$$A\underset{\sim}{u} = \Sigma(\underset{\sim}{u} | \underset{\sim}{e}_i) A \underset{\sim}{e}_i = \Sigma_{i > s} (\underset{\sim}{u} | \underset{\sim}{e}_i) A \underset{\sim}{e}_i$$
$$= \Sigma_{i > s} AP_i \underset{\sim}{u}$$

and so $A = \Sigma_{i > s} AP_i$.

A. 4. Certain norms on $\mathscr{E}$

For all the details connected with the following summary, see Hewitt and Ross [1], Appendix (D. 35) ff.

A. 4. 1. Symmetric norms on $\mathbf{R}^d$. A norm ϕ on $\mathbf{R}^d$ is termed symmetric if and only if it satisfies the following two conditions:

$$\left.\begin{aligned} \phi(\varepsilon_1 x_1, \ldots, \varepsilon_d x_d) &= \phi(x_1, \ldots, x_d), \\ \phi(x_{\pi(1)}, \ldots, x_{\pi(d)}) &= \phi(x_1, \ldots, x_d) \end{aligned}\right\} \qquad (A.\ 4.\ 1)$$

for every $(x_1, \ldots, x_d) \in \mathbf{R}^d$, every choice of the $\varepsilon_i \in \{-1, 1\}$, and every permutation π of $\{1, 2, \ldots, d\}$.

If ϕ is such a norm, the conjugate norm ψ is defined by

$$\psi(y_1, \ldots, y_d) = \sup \Sigma_{i=1}^d x_i y_i \qquad (A.\ 4.\ 2)$$

for every $(y_1, \ldots, y_d) \in \mathbf{R}^d$, the supremum being taken with respect to all $(x_1, \ldots, x_d) \in \mathbf{R}^d$ satisfying $\phi(x_1, \ldots, x_d) \leq 1$.

The conjugate of the conjugate of ϕ is ϕ itself.

To each $p \in [1, \infty]$ corresponds an important symmetric norm ϕ_p on $\mathbf{R}^d$, defined as follows

$$\phi_p(x) = \begin{cases} (\Sigma_{i=1}^d |x_i|^p)^{1/p} & \text{if } p \neq \infty \\ \max_{1 \le i \le d} |x_i| & \text{if } p = \infty. \end{cases} \qquad (A.4.3)$$

The conjugate of ϕ_p is $\phi_{p'}$, where $1/p + 1/p' = 1$. (The usual convention, that $p = \infty$ if and only if $p' = 1$ and vice versa, is adopted here.)

A. 4. 2. Norms on $\mathscr{E}$. Corresponding to each symmetric norm ϕ on $\mathbf{R}^d$ is a norm $\|.\|_\phi$ on $\mathscr{E}$, defined by the formula

$$\|X\|_\phi = \phi(\lambda_1^{\frac{1}{2}}, \ldots, \lambda_d^{\frac{1}{2}}), \qquad (A.4.4)$$

for every $X \in \mathscr{E}$, where $\lambda_1, \ldots, \lambda_d$ denote the eigenvalues of XX^*, repeated each according to its multiplicity. More precisely, suppose that $\zeta \in \mathbf{C} \mapsto P_\zeta$ is the spectral resolution of XX^* (cf. A. 1. 1 and A. 1. 2 above), that $i \mapsto \zeta_i$ is an injection of $\{1, 2, \ldots, r\}$ onto $\sigma(XX^*)$, and that d_i is the dimension of $P_{\zeta_i}(\mathscr{H})$; then $d_1 + \ldots + d_r = d$ and we may take $\lambda_i = \zeta_1$ for $1 \le i \le d_1$, $\lambda_i = \zeta_2$ for $d_1 < i \le d_1 + d_2, \ldots, \lambda_i = \zeta_r$ for $d_1 + \ldots + d_{r-1} < i \le d$.

It then turns out that for every $X \in \mathscr{E}$:

$$\left.\begin{aligned} \|X\|_{\phi_1} &= \operatorname{Tr} |X|, \\ \|X\|_{\phi_2} &= (\operatorname{Tr} XX^*)^{\frac{1}{2}}, \\ \|X\|_{\phi_\infty} &= \|X\|, \\ \|U\|_{\phi_p} &= d^{1/p} \quad \text{if } U \in U(\mathscr{H}). \end{aligned}\right\} \qquad (A.4.5)$$

Of the many properties of the norms $\|.\|_\phi$ on $\mathscr{E}$ we shall need only the following, valid for every $A, B \in \mathscr{E}$:

$$\|A\|_\phi = \sup\{|\operatorname{Tr} AB| : \|B\|_\psi \le 1\}, \qquad (A.4.6)$$

where ψ is the conjugate of ϕ. In particular,

$$|\mathrm{Tr}\, AB| \le \|A\|_\phi \|B\|_\psi\ ; \qquad (A.\,4.\,7)$$

when $\phi = \phi_1$, this generalises (A. 2. 11), and when $\phi = \phi_2$ it generalises (A. 2. 6). Furthermore

$$\|A\|_{\phi_\infty} = \sup\{\ \|AB\|_{\phi_p} : \|B\|_{\phi_p} \le 1\}\ . \qquad (A.\,4.\,8)$$

For proofs of these results, see Hewitt and Ross [1], (D. 39) and (D. 54).

Appendix B

B. 1. Supplement to 2. 6. 1-2. 6. 4

The proof of (2. 6. 1) is indicated in the text.

To obtain (2. 6. 3) from (2. 6. 1), we change x into x^{-1}, so getting

$$\int U(x)^* T U(x) dx = I_U . \operatorname{Tr} T/d(U) .$$

If $(\underset{\sim}{u}_i)$ is an orthonormal base for $\mathscr{H}_U$, this means that

$$\int (U(x)^* T U(x) \underset{\sim}{u}_i | \underset{\sim}{u}_j) dx = \delta_{ij} \operatorname{Tr} T/d(U)$$

or

$$\int (T U(x) \underset{\sim}{u}_i | U(x) \underset{\sim}{u}_j) dx = \delta_{ij} \operatorname{Tr} T/d(U) .$$

Introducing the expansions

$$U(x)\underset{\sim}{u}_i = \Sigma_\alpha (U(x)\underset{\sim}{u}_i | \underset{\sim}{u}_\alpha) \underset{\sim}{u}_\alpha = \Sigma_\alpha \bar{u}_{\alpha i}(x) \underset{\sim}{u}_\alpha ,$$

$$T U(x)\underset{\sim}{u}_i = \Sigma_\alpha \bar{u}_{\alpha i}(x) T \underset{\sim}{u}_\alpha ,$$

$$U(x)\underset{\sim}{u}_j = \Sigma_\beta \bar{u}_{\beta j}(x) \underset{\sim}{u}_\beta ,$$

we obtain

$$\Sigma_\alpha \Sigma_\beta \int \bar{u}_{\alpha i} u_{\beta j} dx . \ (T\underset{\sim}{u}_\alpha | \underset{\sim}{u}_\beta) = \delta_{ij}/d(U) . \Sigma_\gamma (T\underset{\sim}{u}_\gamma | \underset{\sim}{u}_\gamma) .$$

Since T is arbitrary, so are the numbers $(T\underset{\sim}{u}_\alpha | \underset{\sim}{u}_\beta)$, and it follows that

$$\int \bar{u}_{\alpha i} u_{\beta j} dx = \delta_{ij} \delta_{\alpha\beta}/d(U) ,$$

which is equivalent to (2. 6. 3).

The derivation of (2. 6. 4) from (2. 6. 2) is similar.

To derive (2. 6. 9) we proceed as follows. (The proof of (2. 6. 7) is similar and somewhat easier.) Replacing T by FTG* in (2. 6. 1) we get

$$\int U(x)FTG^*U(x)^*dx = I_U . \operatorname{Tr}(FTG^*)/d(U) = \gamma(T)I_U, \text{ say} .$$

Write $F(x) = U(x)F$, $G(x) = U(x)G$, so that

$$\int (F(x)TG(x)^*\underset{\sim}{u}_i | \underset{\sim}{u}_j)dx = \delta_{ij}\gamma(T) .$$

Make expansions of $G(x)^*\underset{\sim}{u}_i$ and $F(x)^*\underset{\sim}{u}_j$ in terms of the $\underset{\sim}{u}_\alpha$:

$$\sum_{\alpha\beta}\sum \int (F(x)\underset{\sim}{u}_\beta | \underset{\sim}{u}_j)\overline{(G(x)\underset{\sim}{u}_\alpha | \underset{\sim}{u}_i)}dx. (T\underset{\sim}{u}_\alpha | \underset{\sim}{u}_\beta) = \delta_{ij}\gamma(T) . \qquad \text{(B. 1. 1)}$$

On the other hand

$$\gamma(T) = d(U)^{-1}\operatorname{Tr}(FTG^*) = d(U)^{-1} \sum_\gamma (FTG^*\underset{\sim}{u}_\gamma | \underset{\sim}{u}_\gamma) ;$$

if we make expansions of $G^*\underset{\sim}{u}_\gamma$, $TG^*\underset{\sim}{u}_\gamma$ and $F^*\underset{\sim}{u}_\gamma$ in terms of the $\underset{\sim}{u}$'s, substitution leads to

$$\gamma(T) = \sum_{\alpha\beta}\sum (T_\alpha | \underset{\sim}{u}_\beta) \sum_\gamma d(U)^{-1}(F\underset{\sim}{u}_\beta | \underset{\sim}{u}_\gamma) \overline{(G\underset{\sim}{u}_\alpha | \underset{\sim}{u}_\gamma)}. \qquad \text{(B. 1. 2)}$$

Since T is arbitrary, (B. 1. 1) and (B. 1. 2) lead to

$$\int (F(x)\underset{\sim}{u}_\beta | \underset{\sim}{u}_j)\overline{(G(x)\underset{\sim}{u}_\alpha | \underset{\sim}{u}_i)}dx = \delta_{ij}d(U)^{-1} \sum_\gamma (F\underset{\sim}{u}_\beta | \underset{\sim}{u}_\gamma)\overline{(G\underset{\sim}{u}_\alpha | \underset{\sim}{u}_\gamma)} .$$

In this we take $\alpha = i$ and $\beta = j$ and then sum over both: the result is

$$\int \operatorname{Tr} F(x). \overline{\operatorname{Tr}G(x)}dx = d(U)^{-1} \sum_{k\gamma}\sum (F\underset{\sim}{u}_k | \underset{\sim}{u}_\gamma)\overline{(G\underset{\sim}{u}_k | \underset{\sim}{u}_\gamma)} .$$

By making expansions, it is verified that the sum on the right is precisely $\operatorname{Tr}(FG^*)$ - see Appendix A, formula (A. 2. 5). Thus we get

$$\int \operatorname{Tr} F(x). \operatorname{Tr}G(x)dx = d(U)^{-1}\operatorname{Tr}(FG^*) ,$$

which is equivalent to (2. 6. 9).

Finally, the proof of (2. 6. 12) comes at once from the expansion

$$h(x) = d(U) \sum_{i\,j}\sum (H\underset{\sim}{u}_i | \underset{\sim}{u}_j)u_{ij}(x)$$

and (2. 6. 3) and (2. 6. 4): these show that

$$(\hat{h}(V)\underset{\sim}{v}_h | \underset{\sim}{v}_k) = d(U) \sum_{i\,j}\sum \int u_{ji}. \bar{v}_{kh}dx. (h\underset{\sim}{u}_i | \underset{\sim}{u}_j) = 0$$

and hence $\hat{h}(V) = 0$ for $V \neq U$; and also, using (2.6.3),

$$(\hat{h}(U)\underset{\sim}{u}_h | \underset{\sim}{u}_k) = d(U) \sum_i \sum_j (H\underset{\sim}{u}_i | \underset{\sim}{u}_j) \int u_{ji} \bar{u}_{kh} \, dx$$
$$= (H\underset{\sim}{u}_h | \underset{\sim}{u}_k) ,$$

so that $\hat{h}(U) = H$.

B. 2. Supplement to 2.7.1.

Note that

$$u_{ij}(x) = (\underset{\sim}{u}_i | U(x)\underset{\sim}{u}_j) = (U(x^{-1})\underset{\sim}{u}_i | \underset{\sim}{u}_j) = \overline{u_{ji}(x^{-1})} ,$$

so that $u_{ij} = \tilde{u}_{ji}$. Also

$$u_{ii}(xy) = (\underset{\sim}{u}_i | U(xy)\underset{\sim}{u}_i) = (U(x^{-1})\underset{\sim}{u}_i | U(y)\underset{\sim}{u}_i)$$
$$= \sum_j (U(x^{-1})\underset{\sim}{u}_i | \underset{\sim}{u}_j) (\underset{\sim}{u}_j | U(y)\underset{\sim}{u}_i)$$

and so

$$u_{ii}(xy) = \Sigma_j \, u_{ij}(y) u_{ji}(x) .$$

Hence

$$\Sigma_i \Sigma_j \, u_{ij}(x) \int f(y)\overline{u_{ij}(y)}dy = \Sigma_i \int \Sigma_j \, f(y) u_{ij}(x)\overline{u_{ij}(y)}dy$$
$$= \Sigma_i \int \Sigma_j \, f(y^{-1}) u_{ij}(x) u_{ji}(y) dy = \Sigma_i \int f(y^{-1}) u_{ii}(yx) dy$$
$$= \int f(y^{-1}) \, \chi_U(yx) dy = \int f(y) \chi_U(y^{-1}x) dy$$
$$= f * \chi_U(x) .$$

Again (Appendix A, formula (A.2.5)):

$$\mathrm{Tr}\, \hat{f}(U)\hat{f}(U)^* = \Sigma_i \Sigma_j |(\hat{f}(U)\underset{\sim}{u}_i | \underset{\sim}{u}_j)|^2$$

and

$$(\hat{f}(U)\underset{\sim}{u}_i | \underset{\sim}{u}_j) = \int f(x). \, (U(x)\underset{\sim}{u}_i | \underset{\sim}{u}_j) dx$$
$$= \int f(x). \, \overline{u_{ji}(x)} dx ,$$

so that

$$\Sigma_i \Sigma_j \left| \int f(x) u_{ij}(x) dx \right|^2 = \mathrm{Tr}\, \hat{f}(U) \hat{f}(U)^* .$$

Bibliography

Banach, S.

[1] Théorie des Opérations Linéaires. Warsaw (1932).

Benedetto, J.

[1] Harmonic analysis on totally disconnected sets. Springer-Verlag, Berlin-Heidelberg-New York (1971).

Bourbaki, N.

[1] Intégration. Ch. 1-6. Act. Sci. et Ind. Nos. 1175, 1244, 1281. Paris (1952, 1956, 1959).

[2] Intégration. Ch. 7, 8. Act. Sci. et Ind. No. 1306. Paris (1963).

[3] Théories Spectrales. Ch. 1, 2. Act. Sci. et Ind. No. 1332. Paris (1967).

Daniell, P. J.

[1] A general form of integral. Ann. of Math. (2) 19 (1917-18), 279-294.

Donoghue, W. F., Jr.

[1] Distributions and Fourier transforms. Academic Press, New York and London (1969).

Drury, S. W.

[1] Sur les ensembles de Sidon. C.R. Acad. Sci. Paris, 271 (1970), 162-3.

Dunford, N. and Schwartz, J. T.

[1] Linear Operators. I. New York (1958).

Edwards, R. E.

[1] A theory of Radon measures on locally compact spaces. Acta Math. 89 (1953), 133-64.

[2] Functional Analysis: Theory and Applications. New York (1965).

[3] Fourier Series: A Modern Introduction. Vols. I, II. New York (1967).

[4] What is the Riemann Integral? Notes on Pure Mathematics 1, A. N. U. (1968).

[5] Paths in Complex Analysis. Notes on Pure Mathematics 4, A. N. U. (1969).

[6] Unbounded integrally positive definite functions. Studia Math. 33 (1969), 185-91.

[7] On functions whose translates are independent. Ann. Inst. Fourier III (1951), 32-72.

Edwards, R. E. and Hewitt, E.

[1] Pointwise limits for sequences of convolution operators. Acta Math. 113 (1965), 181-218.

Edwards, R. E., Hewitt, E. and Ross, K. A.

[1] Lacunarity for compact groups, I. Proc. Nat. Acad. Sci. U. S. A. 68 (1971), 25.

[2] Lacunarity for compact groups, I. To appear Indiana J. Math.

Edwards, R. E. and Price, J. F.

[1] A naively constructive approach to boundedness principles, with applications to harmonic analysis. Enseignement Math. XVI (1971), 255-96.

Ehrenpreis, L.

[1] Fourier analysis in several complex variables. Wiley-Interscience Publishers, New York (1970).

Eymard, P.

[1] L'algèbre de Fourier d'un groupe localement compact. Bull. Soc. Math. France, 92 (1964), 181-236.

Fréchet, M.

[1] Sur les opérations linéaires. I.

[2] Sur les opérations linéaires. II.

[3] Sur les opérations linéaires. III. Trans. Amer. Math. Soc. 5 (1904), 493-9; 6 (1905), 134-40; 8 (1907), 433-46.

Glicksberg, I.

[1] The representation of functionals by integrals. Duke Math. J. k9 (1952), 253-61.

Godement, R.

[1] Les fonctions de type positif et la théorie des groupes. Trans. Amer. Math. Soc. 63 (1948), 1-64.

Hadamard, J.

[1] Sur les opérations fonctionelles. C.R. Acad. Sci. Paris, 136 (1903).

[2] Leçons sur le calcul de variations. Paris (1910).

Halmos, P.

[1] Measure Theory. New York (1950).

[2] Finite Dimensional Vector Spaces. Princeton (1958).

Helly, E.

[1] Über lineare Funktionaloperationen. S.-B. K. Akad. Wiss. Wien Math.-Naturwiss. Kl. 121, IIa (1912), 265-97.

Hewitt, E.

[1] Linear functionals on spaces of continuous functions. Fund. Math. XXXVII (1950), 161-89.

[2] Integration on locally compact spaces. I. Univ. of Washington Pub. in Math. 3 (1952), 71-5.

[3] Integral representations of continuous linear functionals. Ark. f. Mat. (11) 2 (1952), 269-82.

[4] The role of compactness in analysis. Amer. Math. Monthly. 67 (6) (1960), 499-516.

Hewitt, E. and Ross, K. A.

[1] Abstract Harmonic Analysis, Vols. I, II. Berlin (1963, 1970).

Hewitt, E. and Stromberg, K.

[1] Real and Abstract Analysis. Springer-Verlag New York Inc. (1965).

Kakutani, S.

[1] Concrete representations of abstract (M)-spaces. (A characterisation of the space of continuous functions.) Ann. of Math. 42 (1941), 994-1024.

Kahane, J. P.

[1] Some Random series of Functions. Lexington, Mass. (1968).

[2] Séries de Fourier absolument convergentes. Berlin (1970).

Katznelson, Y.

[1] An Introduction to Harmonic Analysis. New York (1968).

Kelley, J. L.

[1] General Topology. New York (1955).

Loomis, L.

[1] Abstract Harmonic Analysis. New York (1953).

Maak, W.

[1] Fastperiodische Funktionen. Berlin (1950).

Macdonald, I. D.

[1] The Theory of Groups. Oxford (1968).

Markov, A.

[1] On mean values and exterior densities. Mat. Sbornik. N. S. 4 (46) (1938), 165-91.

Mayer, R. A.

[1] Summation of Fourier series on compact groups. Amer. J. Math. 89 (1967), 661-92.

Meyer, Y.

[1] Endomorphismes des idéaux fermés de $L^1(G)$, classes de Hardy et séries de Fourier lacunaires. Ann. Sci. Ecole Norm. Sup.

(4) 1 (1968), 499-580.

Naimark, M. A.

[1] Normed rings. Groningen (1959).

Pontryagin, L.

[1] Topological Groups. Princeton (1939).

Radon, J.

[1] Theorie und Anwendungen der absolut additiven Mengenfunktionen. S.-B. K. Akad. Wiss. Wien Math. Naturwiss. Kl. 122 (1913), 1295-438.

Reiter, H.

[1] Classical Harmonic Analysis and Locally Compact Groups. Oxford (1968).

Riesz, F.

[1] Sur les opérations fonctionelles linéaires. C. R. Acad. Sci. Paris, 149 (1909), 974-7.

[2] Sur certains systèmes singuliers d'équations intégrales. Ann. École Norm. Sup. (3) 28 (1911), 33-62.

[3] Démonstration nouvelle d'un théorème concernant les opérations fonctionelles linéaires. Ibid. 3 (31) (1914), 9-14.

[4] Sur la représentation de opérations fonctionelles linéaires par des intégrales de Stieltjes. Proc. Roy. Physiog. Soc. Lund, 21, n. 16 (1952), 145-51.

Riss, J.

[1] Éléments de calcul différentiel et théorie des distributions sur les groupes abéliens localement compacts. Acta. Math. 89 (1953), 45-105.

Rudin, J.

[1] Fourier Analysis on Groups. New York (1962).

Saks, S.

[1] Theory of the Integral. 2nd ed. Warsaw (1937).

[2] Integration in abstract metric spaces. Duke Math. J. 4 (1938), 408-11.

Stone, M. H.

[1] Notes on integration. I.

[2] Notes on integration

[3] Notes on integration. III.

[4] Notes on integration. IV.

Proc. Nat. Acad. Sci. U.S.A. 34 (1948), 336-42, 447-55, 483-90; 35 (1949), 50-8.

Varadarajan, V. S.

[1] On a theorem of F. Riesz concerning the form of linear functionals. Fund. Math. XLVI (1958), 209-20.

Volterra, V.

[1] Theory of Functionals. London (1931).

Weil, A.

[1] L'intégration dans les groupes topologiques et ses applications. Paris (1953).

Wigner, E.

[1] Group theory. New York (1959).

ADDENDUM

Newman, Morris

Matrix Representations of Groups. NBS Applied Maths. Series, 60. Washington, D.C. (1968).

Maurin, K.

General eigenfunction expansions and unitary representations of topological groups. Monografie Mat., Tom 48, Warsaw (1968).